Biology of Extracellular Matrix

For further volumes:
www.springer.com/series/8422

Mauro S.G. Pavão
Editor

Glycans in Diseases and Therapeutics

Springer

Editor
Mauro S.G. Pavão
Programa de Glicobiologia
Instituto de Biquímica Médica and Hospital
Universitário Clementino Fraga Filho
Universidade Federal do Rio de Janeiro
Av. Prof. Rodolpho Paulo Rocco, 255
Cidade Universitaria – Rio de Janeiro
CEP 21941-913 - Rio de Janeiro
Brazil
mpavao@hucff.ufrj.br

ISSN 0887-3224 e-ISSN 2191-1959
ISBN 978-3-642-16832-1 e-ISBN 978-3-642-16833-8
DOI 10.1007/978-3-642-16833-8
Springer Heidelberg Dordrecht London New York

Library of Congress Control Number: 2011925996

Cover design: deblik

Printed on acid-free paper

Springer is part of Springer Science+Business Media (www.springer.com)

Foreword

I am delighted to provide a foreword to "Glycans in Diseases and Therapeutics" edited by Dr. Mauro Pavao, at his invitation. Although I do not have as much knowledge of all the glycans highlighted in this book as is necessary to introduce the recent advances in individual glycan fields, I found the book intriguing as both a glycobiologist and a structural biochemist, and useful not only for researchers but also for graduate students to learn about the rapidly growing molecular fields, which were ignored for decades, but are now recognized as essential to understand the numerous critical biological phenomena in development or life processes.

I believe that the structure of polysaccharides from marine invertebrates analogous to that of mammalian heparinoids turned Dr. Pavao's scientific interests to the structure–activity relationship and the possibility of therapeutic applications, almost 20 years ago. I have always thoroughly enjoyed his work, which is often impressive since he provides solid structural data and describes interesting activities of complicated molecules. I think, I first met him at one of the Gordon Conferences on Proteoglycans in 2000. I have enjoyed the discussions with him almost every year since then at the same and other conferences on proteoglycans.

Over the years, his interests have been turned to various pharmacological activities of glycans such as antimetastasis, antithrombosis, and anti-inflammation. This book is unique that it was designed to encompass the most recent advances in biological and pharmacological research on various glycans including not only heparan sulfate and chondroitin/dermatan sulfate glycosaminoglycans (GAGs), but also *N*- and *O*-glycans of glycoproteins. Dr. Pavao invited young independent frontier researchers, who are at the cutting-edge in each research field, to be the authors.

When I entered this field, almost 35 years ago, it was small and relatively inactive. However, it is now clear, based on recent accumulating and direct evidence, that glycans are essential components for life. It is now timely to compile review articles of representative sections of the most interesting studies. I express my respect for Dr. Pavao for his contribution to the field as an editor, and to all the authors for their efforts.

Hokkaido University, Hokkaido, Japan   Kazuyki Sugahara

Preface

Carbohydrates are a major component of the extracellular matrix (ECM) where they associate with proteins to form glycoproteins or proteoglycans, or exist as long-chain disaccharides (e.g., hyaluronic acid). All ECM proteins except elastin have associated sugar, and, in some cases, ECM proteins require proper glycosylation to achieve full biological activity. It is also now clear that many ECM proteins have carbohydrate-binding domains that specifically recognize and interact with glycoconjugates on other matrix components and on the cell surface. Carbohydrates have been implicated in a wide variety of processes, ranging from cell adhesion and migration to matrix assembly, growth factor sequestration and regulation, involvement in many aspects of immune function, binding of plasma proteins, and control of thrombogenesis. Proteoglycans and other glycoconjugates were reviewed in earlier volumes in this series (*Cell Surface and Extracellular Glycoconjugates: Structure and Function*, David D. Roberts and Robert P. Mecham, editors; and *Biology of Proteoglycans*, Thomas N. Wight and Robert P. Mecham, editors). This volume reviews the most recent findings on the role of glycans in the development of diseases and the possible therapeutic use of this class of molecules. It shows how the interaction of glycans with growth factors, growth factor-binding proteins, extracellular proteases, protease inhibitors, chemokines, morphogens, and adhesive proteins regulates inflammation, infection, cancer, atherosclerosis, thrombosis, and embryonic stem cell biology. Further, an extensive survey of the structure and pharmacological effects of unique marine GAGs, and also the possibility to use these glycans as therapeutic agents are discussed.

Heparan sulfate is a linear polysaccharide composed of glucosamine and uronic acid (glucuronic acid or iduronic acid) disaccharide repeats with various types of sulfation modifications. In tissues, heparan sulfate covalently attaches to core proteins to form heparan sulfate proteoglycans (HSPG), and are abundant at the cell surface and in the ECM. In Chap. 1, Wang reviews our current understanding of the cellular and molecular mechanisms of heparan sulfate in the regulation of inflammation and angiogenesis. This chapter focuses on the regulatory roles of heparan sulfate on key inflammatory molecules and on the vascular endothelial growth factor (VEGF), the master proangiogenic factor of angiogenesis. For most intracellular pathogens, cell surface HSPGs serve as a scaffold that facilitates the interaction of microbes with secondary receptors that mediate host cell entry. Consistent with this mechanism, addition of heparan sulfate (HS) or its pharmaceutical functional mimic, heparin, inhibits microbial attachment and entry into cultured host cells, and HS-binding pathogens can no longer attach or enter cultured host cells whose HS expression has been reduced by enzymatic treatment or chemical mutagenesis. In Chap. 2, Bartlett and Park provide a mechanistic overview of our current understanding of how certain microbial pathogens subvert HSPGs to promote their infection, using specific HSPG-pathogen interactions as representative examples.

Alteration in cellular glycosylation is a common phenotypic change associated with malignant transformation and cancer progression. Cell surface oligosaccharides carried on glycoproteins and glycolipids mediate communication among cells and facilitate cell adhesion, processes that are central during cancer progression. Accumulating evidence indicates that glycans contribute to tumor invasion, metastasis, and angiogenesis. Borsig (Chap. 3) provides an overview on cancer-specific changes of glycosylation on *O*- and *N*-glycans, with the focus on the function of these oligosaccharides in cancer progression.

Chapter 4 by Vicente, Godoy and Werneck focuses on the roles of different GAGs located in the vessel wall in the pathogenesis of atherosclerosis and thrombosis. GAGs in atherosclerosis can help to regulate atherogenesis through their ability to retain lipoproteins in the vessel wall. Prolonged retention of lipoproteins may render them susceptible to chemical modifications, leading to their aggregation, cellular uptake, and lipid accumulation. GAGs can also act as anticoagulant molecules because of their ability to interact with anticoagulant proteins like antithrombin and heparin cofactor II, promoting their activation and increasing their ability to inhibit thrombin.

In Chap. 5, we return to heparan sulfate where Pickford, Holley, Meade, and Merry discuss the role of this important glycoconjugate in early development. Mouse embryonic stem (ES) cells produce a poorly sulfated HS that may protect from prodifferentiation cues (e.g., fibroblast growth factors), and HS epitopes can identify cells with hemangioblast potential. Therefore, HS-sequences can identify functionally unique populations of cells and so have potential applications in the development of cell-replacement therapies for degenerative conditions. How HS chains influence differentiation events is unclear, but possible mechanisms include interactions of heparan with cytokines and chemokines. Indeed, a well-known function of HS is its ability to serve as a coreceptor to modulate cell growth, survival, and movement. These functions become relevant to heparan's role in disease as well as development. In Chap. 6, Gassar, Ibrahim, and Götte discuss the role of HS in tumor progression and cancer therapy. As matrix receptors, HSPGs act in concert with integrins to regulate tumor cell motility. As binding partners for matrix metalloproteinases and protease inhibitors, they regulate the proteolytic microenvironment of tumors, thus modulating metastatic spread. The development of glycan-based drugs targeting these biological functions has become an area of intense research in cancer biology. This chapter discusses some of the promising results that have been obtained both in animal models and in clinical trials.

The final chapter by Kozlowski, Gomes, Silva, Pereira, Silva, and Pavão focuses on the interesting sulfated GAGs present on marine organisms. These glycoconjugates possess pharmacological properties ranging from anticoagulant and antithrombotic to antimetastatic and anti-inflammatory. In this chapter, the authors review the phylogenetic distribution, the structure, and the biological effects of the marine GAGs, as well as the molecular mechanisms involved in some of their biological activities. The possibility to use these glycans as therapeutic agents is also discussed.

It is hoped that the information in this volume will provide useful information to the research community about the broad range of functions associated with glycoconjugates in the context of development and disease. This field is especially exciting to follow because it impacts basic, applied, and clinical aspects of biomedical research. It is clear that additional practical applications will be forthcoming with advances in our basic knowledge of glycans in disease.

Rio de Janeiro, Brazil Mauro S.G. Pavão

Contents

1 **Heparan Sulfate Proteoglycan in Inflammation and Angiogenesis** 1
Lianchun Wang

2 **Heparan Sulfate Proteoglycans in Infection** 31
Allison H. Bartlett and Pyong Woo Park

3 **Glycans in Cancer** 63
Lubor Borsig

4 **Glycosaminoglycans in Atherosclerosis and Thrombosis** 83
C.P. Vicente, J.A.P. Godoy, and C.C. Werneck

5 **Glycans in Embryonic Stem Cells** 113
Claire Pickford, Rebecca Holley, Kate Meade, and Catherine Merry

6 **Heparan Sulfate Proteoglycans in Cancer Therapy** 139
Ezeddin Salem Gassar, Sherif A. Ibrahim, and Martin Götte

7 **Structure and Biological Activities of Glycosaminoglycan Analogs from Marine Invertebrates: New Therapeutic Agents?** 159
Eliene O. Kozlowski, Angélica M. Gomes, Christiane Sobral Silva, Mariana Sá Pereira, Ana Cristina E.S. de Vilela Silva, and Mauro S.G. Pavão

Index 185

Chapter 1
Heparan Sulfate Proteoglycan in Inflammation and Angiogenesis

Lianchun Wang

Abstract Heparan sulfate is a linear polysaccharide composed of glucosamine and uronic acid (glucuronic acid or iduronic acid) disaccharide repeats with various types of sulfation modifications. More than 26 enzymes participate in the biosynthesis of heparan sulfate, which includes two major steps of chain polymerization and chain modification. The chain modification steps proceed sequentially but are incomplete, resulting in enormous structural diversity in mature heparan sulfate. In tissues, heparan sulfate covalently attaches to core proteins to form heparan sulfate proteoglycans, and are abundant at the cell surface and in the extracellular matrix. Studies have demonstrated that heparan sulfate interacts with growth factors, growth factor binding proteins, extracellular proteases, protease inhibitors, chemokines, morphogens, and adhesive proteins to critically regulate cell functions under both physiological and pathological conditions. In this chapter, I will review our current understanding of the cellular and molecular mechanisms, and the structure–function relationship of heparan sulfate in the regulation of inflammation and angiogenesis, with particular focus on the regulatory roles of heparan sulfate on the key inflammatory molecules, selectin and chemokine, and on the vascular endothelial growth factor, the master proangiogenic factor of angiogenesis.

1.1 Introduction

Heparan sulfate and heparin, a highly sulfated form of heparan sulfate, are linear polysaccharides composed of glucosamine and uronic acid (glucuronic acid or iduronic acid) disaccharide repeats with various types of sulfation modifications (Bernfield et al. 1999; Esko and Selleck 2002). Heparan sulfate covalently attaches to core proteins in tissues to form heparan sulfate proteoglycans (HSPGs) and are

---

L. Wang
Complex Carbohydrate Research Center, and Department of Biochemistry and Molecular Biology, University of Georgia, 315 Riverbend Road, Athens, GA 30602, USA
e-mail: lwang@ccrc.uga.edu

M.S.G. Pavão (ed.), *Glycans in Diseases and Therapeutics*, Biology of Extracellular Matrix,
DOI 10.1007/978-3-642-16833-8_1,

abundant at the cell surface and in the extracellular matrix (Bernfield et al. 1999; Esko and Selleck 2002). Heparin is restricted in mast cells of connective tissue and is enriched in liver, lung, intestine, and skin (Forsberg et al. 1999; Bishop et al. 2007). Although heparin has been used clinically as a potent anticoagulant for more than 70 years, the physiological function of endogenous heparin appears to be unrelated to blood coagulation. Instead, heparin is essentially required for protease storage in the secretory granules in mast cells (Forsberg et al. 1999). The biological functions of heparan sulfate have also gradually been recognized and been shown to critically regulate various aspects of cell functioning under both physiological and pathological conditions by interacting with growth factors, growth factor binding proteins, extracellular proteases, protease inhibitors, chemokines, morphogens, and adhesive proteins (Bernfield et al. 1999; Bishop et al. 2007). Intriguingly, studies have further emerged showing that the regulatory roles of heparan sulfate occur in a spatio-temporal manner and are implicated in distinct arrays of molecular pathways during different biological processes (Allen et al. 2001; Ledin et al. 2004). For better understanding of the structure, biosynthesis and general biological functions of heparan sulfate and HSPGs, readers are referred to other chapters in this book as well as excellent reviews in the literature (Bernfield et al. 1999; Esko and Selleck 2002; Bishop et al. 2007; Hacker et al. 2005; Beauvais and Rapraeger 2004; Gotte 2003; Sasisekharan et al. 2002; Iozzo and San Antonio 2001; Gallagher 2001; Forsberg and Kjellen 2001; Esko and Lindahl 2001; Park et al. 2000b; Varki and Varki 2002; Parish 2006). In this chapter, the recent progress in understanding the cellular and molecular mechanisms and the structure–function relationship of heparan sulfate and heparin in regulating inflammation and angiogenesis will be reviewed, with a particular focus on their interactions with the key inflammatory molecules, selectin and chemokine, and with the vascular endothelial growth factor (VEGF), the master proangiogenic factor of angiogenesis (Wang et al. 2002, 2005; Fuster et al. 2007).

1.2 The Anti-inflammatory Function of Heparin

1.2.1 Leukocyte Trafficking and Inflammation

An essential feature of the inflammatory response is the rapid recruitment of leukocytes from the blood to the site of inflammation, usually through postcapillary venules (Parish 2006; Springer 1994). This recruitment occurs as a multistep process, including the initial attachment and rolling of leukocytes on the inflamed endothelium, activation of leukocytes by endothelial cell-bound chemokines, stable adherence of the activated leukocytes to the endothelium, degradation of the subendothelial basement membrane, and migration of leukocytes along chemokine gradients into the target tissue. The recruited leukocytes at the target tissues facilitate host responses to tissue injury and pathogen invasion (Springer 1994).

The complex leukocyte recruitment process is mediated and coordinated by diverse and multiple molecules at different stages. Initial tethering and rolling of leukocytes on the inflamed endothelium are mediated by the selectin family members, including lymphocyte (L)-, platelet (P)- and endothelial-cell (E)-selectin, and their counter ligands (Springer 1994; Norgard-Sumnicht et al. 1993). L-selectin is constitutively expressed by leukocytes, and P- and E-selectin are expressed on endothelial cells activated by proinflammatory cytokines. P-selectin is rapidly mobilized to the surface of endothelial cells exposed to thrombin or histamine, whereas the expression of E-selection by endothelial cells appears several hours after receiving the inflammatory stimuli, such as IL-1, TNF-α and endotoxin. Therefore, P-selectin and L-selectin appear to mediate the initial events in leukocyte infiltration due to their rapid appearance on activated endothelium and constitutive expression on leukocytes, respectively (Springer 1994; Parish 2006). Blockage of the L- and P-selectin-mediated leukocyte tethering and rolling in blood vessels represents a reasonable approach to inhibit unwanted inflammatory responses (Parish 2006; Springer 1994; Ebnet et al. 1996; Koenig et al. 1998; Wang et al. 2002).

1.2.2 Heparin Inhibits Inflammation Mainly Through Blockage of L- and P-Selectin-Mediated Leukocyte Recruitment

All three selectins contain an amino terminal, a calcium-dependent carbohydrate recognition domain that binds to sialylated, fucosylated carbohydrate antigens related to sialyl Lewis^x [SLe^x, Neu5Acα2,3Galβ1,4(Fucα1,3)GlcNAcβ-] (Varki 1994). The SLe^x determinants present on specific selectin ligands, such as P-selectin glycoprotein ligand-1, E-selectin ligand-1 and glycosylation-dependent cell adhesion molecule-1 (Varki 1994; Leppanen et al. 2000). The binding of selectin to these ligands appears to involve sulfate residues either on the carbohydrate moiety or on the protein as sulfated tyrosine residues in proximity to the carbohydrate chain (Hemmerich et al. 2001; Leppanen et al. 2000). Heparin, a highly sulfated form of heparan sulfate and a classic anticoagulant, was found to bind L- and P-selectin but not E-selectin, and to inhibit the binding of L- and P-selectin to SLe^x-related compounds and to SLe^x determinants on HL-60 cells (Norgard-Sumnicht et al. 1993; Nelson et al. 1993; Norgard-Sumnicht and Varki 1995; Giuffre et al. 1997; Koenig et al. 1998). Furthermore, injected heparin inhibits inflammatory responses in vivo (Tyrrell et al. 1999; Nelson et al. 1993). In our studies, we observed that heparin inhibits the binding of L- and P-selectin to immobilized SLe^x and the adhesions of SLe^x-bearing U937 cells to immobilized P- and L-selectin and to P-selectin-expressing endothelial cells (Table 1.1) (Wang et al. 2002). These observations validated and also extended previous findings of others, and suggested that the anti-inflammatory effects of heparin might be due to a blockage of P- and L-selectin-mediated leukocyte recruitment. This idea was tested by examining the

Table 1.1 Anticoagulant activity of heparin and its chemically modified derivatives

Heparinoids	Anti-Xa activity (units/mg)	Inhibition of selectin–PAA–sLe^x binding (IC_{50}, μg/ml)		Inhibition of cell-selectin adhesion (IC_{50}, μg/ml)		Inhibition of cell–cell adhesion (IC_{50}, μg/ml)
		L-selectin	P-selectin	L-selectin	P-selectin	
Heparin	149	0.1	2.5	0.4	0.3	0.5
Oversulfated heparin (OS)	12.5	0.01	1.8	0.1	0.02	0.08
2-O, 3-O desulfated heparin (2/3DS)	1.7	0.4	20	3	4	2
N-desulfated/*N*-acetylated heparin (NDS)	0.1	1	10	0.7	0.9	25
N-,2-,3-*O*-desulfated heparin (N/2/3DS)	0.1	4	100	5	8	50
6-*O*-desulfated heparin (6DS)	0.1	85	450	25	30	350
Carboxyl reduced heparin (CR)	0.1	2	10	3	8	50

The anticoagulant activity of the heparinoids was analyzed by aminolytic antifactor Xa assay. All the raw data were converted into percentages for comparative purposes using the formula: % of maximum = [(average of duplicates) − (negative control)]/[(positive control) − (negative control)] × 100. IC_{50} values represent heparinoid concentrations that achieve 50% inhibition in the corresponding assays

anti-inflammatory effect of heparin in L- and P-selectin single knockout mice as well as in P/L-selectin double knockout mice in vivo, including the inflammatory models of thioglycollate-induced acute peritonitis and allergic (delayed-type hypersensitivity) contact dermatitis (Wang et al. 2002). It was observed that, compared to wildtype mice, the L- and P-selectin single knockout mice both showed impaired inflammation in the two models (Fig. 1.1b), indicating that the inflammatory responses in the two models are mediated by both L- and P-selectin. The two models are appropriate to use in testing whether the anti-inflammatory effect of heparin in vivo is due to the blockage of L- and P-selectin-mediated leukocyte recruitment. Compared to wildtype mice, the inflammatory responses in the L- and P-selectin single knockout mice were attenuated, and heparin treatment further extended this attenuation (Fig. 1.1b). However, the extended attenuation was not seen in P/L-selectin double-deficient mice (Fig. 1.1b). The enhanced attenuated inflammatory responses and lack of an additional attenuation by heparin treatment in the P/L-selectin double-deficient mice further confirmed that the inflammatory responses in both models were both P- and L-selectin-dependent, and also indicated that the anti-inflammatory effect of heparin was achieved primarily by blocking both P- and L-selectin functions (Wang et al. 2002).

Heparin has been known to interact with multiple components of the inflammation cascade, including integrins, cytokines, neutrophil-derived elastases, complement activation, and platelet-activating factors (Tyrrell et al. 1999). However, in our studies we demonstrated that the in vivo anti-inflammatory effect of heparin was mainly achieved by blocking L- and P-selectin functions. This might be explained by the fact that P- and L-selectin-mediated interactions are the necessary first step in the

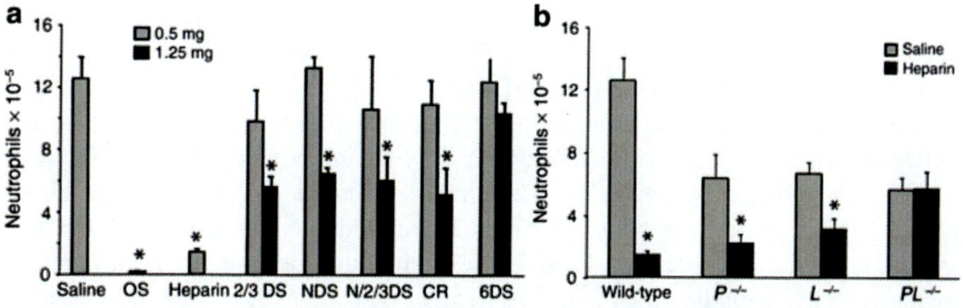

Fig. 1.1 Heparin inhibits thioglycollate-induced acute peritonitis. (**a**) Heparin and heparinoids at 0.5 (*gray bar*) or 1.25 mg (*black bar*) per wildtype mouse were given intravenously 5 min after thioglycollate intraperitoneal injection. The number of neutrophils recruited in the peritoneal cavity was quantified. *Significant statistical difference when comparison was made with heparin treatment. (**b**) Thioglycollate-induced acute peritonitis was elicited in wildtype C57BL/6 and selectin-deficient mice, and the mice received heparin treatment at 0.5 mg/mouse as described in (**a**). *Significant statistical difference when comparison was made between heparin and saline treatment. $P^{-/-}$, P-selectin knockout mice; $L^{-/-}$, L-selectin knockout mice; $PL^{-/-}$, P- and L-selectin double knockout mice

inflammatory cascade, occurring prior to the participation of integrins, cytokines, proteases, and to chemokine presentation and gradient-directed leukocyte migration. Thus, the potential inhibitory effects of heparin on these later steps of inflammatory response might occur, but are not reflected in our inflammatory models.

1.2.3 The Anti-inflammatory Effect of Heparin Does Not Correlate with Its Anticoagulant Activity and Critically Depends on the Presence of 6-O-Sulfate of Glucosamine Residues

Heparin consists of repeating disaccharide units containing D-glucuronic acid (GlcA) or L-iduronic acid (IdoA), and a glucosamine residue that is *N*-sulfated (GlcNS), *N*-acetylated (GlcNAc), or occasionally unsubstituted (GlcNH2). The disaccharides may be further sulfated at C6 or C3 in the glucosamine residues and at C2 in the uronic acid residues. The potent anticoagulant activity of heparin depends on a specific arrangement of sulfated sugar units and uronic acid epimers encompassing $GlcA_{NS6S}$–IdoA–$GlcA_{NS3S6S}$–$IdoA_{2S}$–$GlcA_{NS6S}$, which form a binding site for antithrombin (Esko and Selleck 2002). Much less information is available about the specific oligosaccharide structures in heparin that interact with P- and L-selectin (Nelson et al. 1993; Norgard-Sumnicht et al. 1993; Norgard-Sumnicht and Varki 1995; Koenig et al. 1998; Giuffre et al. 1997; Borsig et al. 2001). Thus, we sought to understand the relationship of heparin fine structure in its interaction with L- and P-selectin. Using heparin as the parent compound and various chemical approaches, several heparinoids were generated, including oversulfated heparin (OS-heparin), *N*-desulfated/*N*-acetylated heparin (NDS-heparin), 2-*O*, 3-*O*-desulfated heparin (2/3DS-heparin), 6-*O*-desulfated heparin (6DS-heparin), and carboxyl-reduced heparin (CR-heparin) (Fig. 1.2). As expected, the desulfation as well as the reduction in heparin carboxyl groups resulted in the heparinoids losing anticoagulant activity, which was determined by the aminolytic antifactor Xa assay (Table 1.1) (Wang et al. 2002). To determine the structure– function relationship of heparin and its anti-inflammatory effect, the heparinoids were initially tested for their inhibition on the interaction of L- and P-selectin with SLe^x among various experimental systems, including the binding of L- and P-selectin to SLe^x, and SLe^x-bearing U937 cell adhesions to immobilized P- and L-selectins and to P-selectin expressing endothelial cells. It was observed that, compared to heparin, the OS-heparin showed an enhanced blocking effect, whereas all the desulfated-heparins and the CR-heparin exhibited reduced inhibitory roles (Table 1.1). Intriguingly, among the desulfated-heparins and the CR-heparin, 6DS-heparin showed the least blocking effect, which was about 4–45 times lower than others. In heparin, *N*-sulfate is the most abundant sulfation modification. Therefore, these observations indicate that the dramatic loss in the capacity of 6DS-heparin to bind to L- and P-selectin is not due to a reduction in the negative charge of the molecule, but is, instead, mainly due to the removal of the

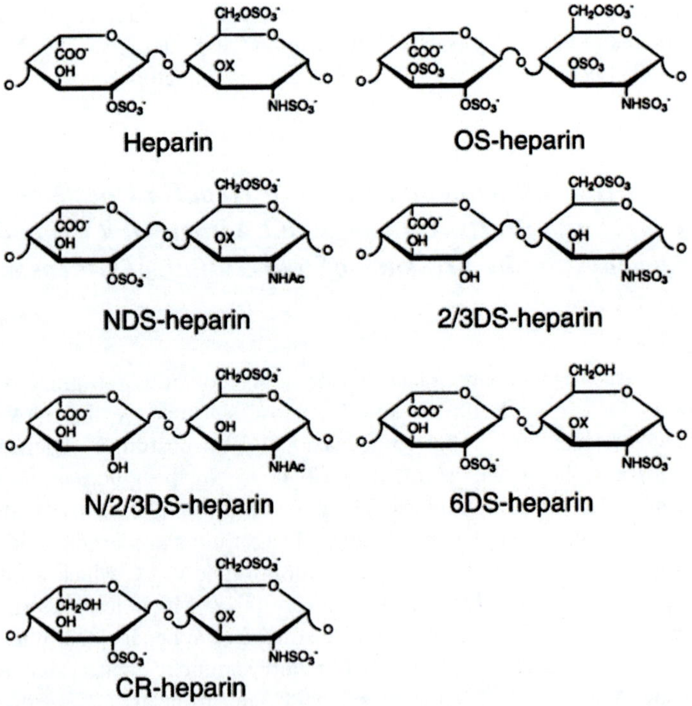

Fig. 1.2 Representative disaccharide units of heparin and chemically modified heparinoids. Each disaccharide illustrates a characteristic unit in the indicated preparation and does not represent the overall structure of the chains. X: H or SO_3^-

6-*O*-sulfate in GlcN residues, thus revealing that the 6-*O*-sulfate of heparin plays a dominant role in determining heparin's interaction with L- and P-selectin. This observation also suggests that the 6-*O*-sulfates of GlcN residues may be critically required for heparin's anti-inflammatory effect in vivo. This was further tested in the mouse inflammatory models of thioglycollate-induced acute peritonitis and allergic contact dermatitis. The in vivo studies recapitulated the in vitro results (Fig. 1.1a), and demonstrated that the anti-inflammatory effect of heparin critically requires the presence of the 6-*O*-sulfates in GlcN residues. Furthermore, with the exception of 6DS-heparin, all other heparinoids retained potent anti-inflammatory effects, even though they lacked anticoagulant activities (Table 1.1; Fig. 1.1a). Therefore, these observations also revealed that the anticoagulant activity of heparin is not required for its anti-inflammatory effect. Together, these structure–function relationship studies laid out a structural basis for the development of non-anticoagulant heparin as potent anti-inflammatory drugs, which can be administrated at high concentrations without fear of the unwanted bleeding side effects that occur with the clinically used heparin.

In summary, these studies demonstrate for the first time that the anti-inflammatory effect of heparin is achieved mainly by blocking the L- and P-selectin-mediated leukocyte recruitment. The structure–function studies further revealed that the 6-*O*-sulfate groups of GlcN residues are critically required for heparin's anti-inflammatory effect and some non-anticoagulant heparin may be developed as potent, novel anti-inflammatory drugs.

1.3 The Role of Heparan Sulfate in Leukocyte Recruitment in Inflammation

Studies have shown that heparin efficiently blocks inflammation in vivo, but heparin is rarely found in blood vessels where L- and P-selectin-mediated leukocyte recruitment occurs during inflammation. Instead, heparan sulfate is abundantly present on the vascular cell surface and in the perivascular matrix, particularly in the basement membrane, suggesting that heparan sulfate may mimic heparin's function to participate in inflammatory response in vivo. Indeed, biochemical and cell-based in vitro studies established that endothelial heparan sulfate is a potential ligand for P- and L-selectin (Norgard-Sumnicht et al. 1993; Norgard-Sumnicht and Varki 1995; Koenig et al. 1998; Giuffre et al. 1997), and can bind proinflammatory chemokines (Witt and Lander 1994; Luster et al. 1995; Hoogewerf et al. 1997), transport chemokine across endothelial cells and present chemokine to leukocytes (Middleton et al. 1997). However, the precise role of heparan sulfate during the inflammatory response in vivo was unclear. In our recent studies we generated heparan sulfate-deficient mice and specifically examined the role of endothelial and leukocyte heparan sulfate in inflammation in vivo (Wang et al. 2005). These studies revealed that endothelial heparan sulfate essentially modulates multiple stages of the inflammatory response (Wang et al. 2005).

1.3.1 Generation of Endothelial- and Leukocyte-Specific Ndst1-Deficient Mice

To examine the function of heparan sulfate in inflammation in vivo, the Cre-loxP system was strategized to ablate *N*-acetylglucosamine *N*-deacetylase–*N*-sulfotransferases (Ndst), the genes essential for heparan sulfate biosynthesis, in mouse endothelial cells and leukocytes. Mice have four Ndst genes (Aikawa et al. 2001). In endothelial cells, only Ndst1 and Ndst2 are abundantly expressed (Wang et al. 2005). Ndst2 null mice are viable and do not display any alteration in heparan sulfate biosynthesis except in mast cells, resulting in defective granule formation (Forsberg et al. 1999). In contrast, Ndst1 null mice show abnormal lung and forebrain development, die shortly after birth and have decreased sulfation of heparan sulfate in all tissues examined (Ringvall et al. 2000; Grobe et al. 2002), suggesting that Ndst1 plays a dominant role in the *N*-sulfation modification of endothelial and leukocyte heparan sulfates. Therefore, we targeted Ndst1 and generated a conditional allele of Ndst1 (assigned as "$Ndst1^{f}$" here) by placing loxP recombination sites flanking exon 2, which encompasses part of the 5′- untranslated region, the start codon and sequence encoding 171 amino acids of the enzyme, including part of the *N*-deacetylase domain (Wang et al. 2005). We established two lines of mice homozygous for the $Ndst1^{f}$ allele from two independently generated embryonic stem cell lines. The incorporation of the LoxP sites into the mouse genome did not have deleterious effects on the survival of the engineered mice, as evidenced by the fact that the $Ndst1^{f/f}$ mice appear normal and the interbreeding of $Ndst1^{f/+}$ mice yielded offspring at the expected Mendelian ratio (27% $Ndst1^{+/+}$, 49% $Ndst1^{f/+}$ and 24% $Ndst1^{f/f}$, $n = 146$). To selectively inactivate Ndst1 in endothelial cells and leukocytes, the $Ndst1^{f/f}$ mice were bred with transgenic mice bearing Cre recombinase under the control of the Tek promoter (Kisanuki et al. 2001). The resulting $TekCre^{+}Ndst1^{f/f}$ (assigned as "Ndst1 mutant" here) mice thrived, reproduced normally, and were indistinguishable in health and behavior from their $Tie2Cre^{-}Ndst1^{f/f}$ (assigned as "wildtype" here) littermates. To determine whether Ndst1 was efficiently ablated in the mutant mice, Southern blot analysis was performed and showed that more than 95% of the $Ndst1^{f}$ were inactivated in endothelial cells and leukocytes in the Ndst1 mutant mice (Wang et al. 2005). To determine whether the Ndst1 ablation alters heparan sulfate structure, heparan sulfate from primary endothelial cells and leukocytes of the mutant mice was analyzed by *N*-sulfation and disaccharide composition analyses, and showed 50–60% reduction in *N*-, 2-*O* and 6-*O*-sulfation (Fig. 1.3a, b) (Wang et al. 2005; Fuster et al. 2007), which is consistent with the known coupling of the *O*-sulfation modification with the *N*-sulfation reaction. To confirm that the heparan sulfate structural alteration in isolated cells was manifested in tissues, fixed mouse tissues were stained with biotinylated FGF-2, which requires both *N*- and 2-*O*-sulfation for binding to heparan sulfate (Ishihara et al. 1993; Bai and Esko 1996). We observed that the binding of FGF-2 to high endothelial venules of the mutant lymph nodes was considerably reduced (Fig. 1.3c, d), which is consistent with our

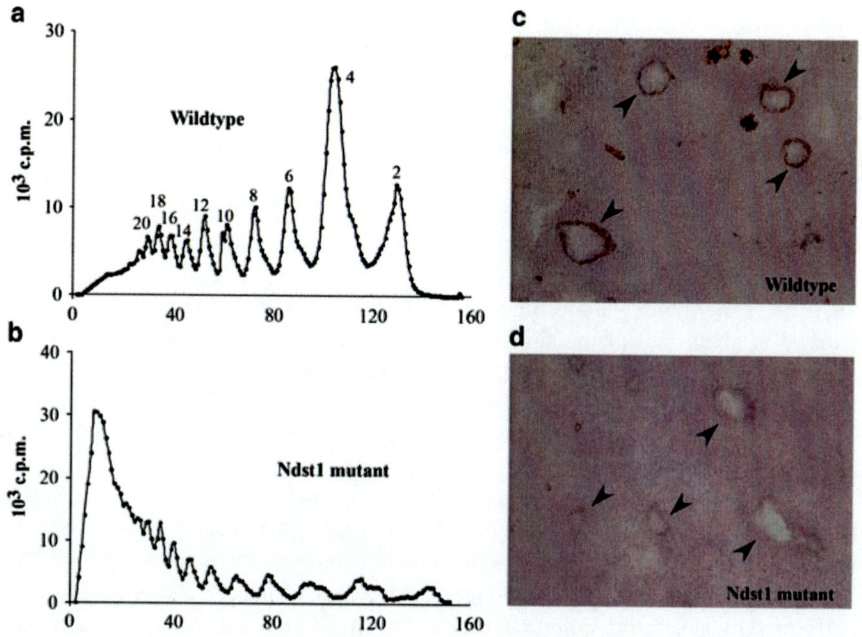

Fig. 1.3 Ndst1 ablation reduces sulfation modification of endothelial heparan sulfate. (**a**, **b**) *N*-sulfation analysis. Isolated endothelial heparan sulfate was cleaved with nitrous acid at a pH of 1.5. The resulting oligosaccharides were differentiated by size chromatography. The "areas under the peaks" were used to estimate the degrees of glucosamine *N*-sulfation. The wildtype endothelial heparan sulfate contained about 45% sulfated glucosamine residues, whereas this value was reduced to 15% in Ndst1 mutant endothelial heparan sulfate. (**c**, **d**) Tissue sections of mouse lung lymph node were stained with FGF-2, which requires both *N*- and 2-*O*-sulfate for high affinity binding. *Arrowheads* point to heparan sulfate-rich basement membranes underlying endothelial cells

in vitro structural analysis (Fig. 1.3a, b). Taken together, these characterizations demonstrate that Ndst1 deficiency altered heparan sulfate structure bearing on both endothelial cells and leukocytes in vivo, and the mutant mice present an ideal test system to determine the biological function of endothelial and leukocyte heparan sulfate in vivo.

1.3.2 *Endothelial Heparan Sulfate Critically Modulates Inflammatory Responses In Vivo*

To examine the effect of altering endothelial and leukocyte heparan sulfate on inflammation, thioglycollate-induced acute peritonitis was induced in the Ndst1 mutant and their littermate wildtype control mice. About 40% fewer neutrophils infiltrated into the peritoneal cavities of the Ndst1 mutant than that of the wildtype control mice ($4.3 \times 10^6 \pm 1.2 \times 10^6$ versus $7.5 \times 10^6 \pm 1.6 \times 10^6$ cells, respectively; $P < 0.05$; Fig. 1.4a). To determine whether other types of inflammation were similarly affected, the Ndst1 mutant mice were further examined by the allergic contract dermatitis model by sensitizing and challenging the mice with hapton oxazolone. The Ndst1 mutant mice showed 32% less ear thickening than that of wildtype mice (0.24 ± 0.02 versus 0.36 ± 0.03 mm, respectively, $P < 0.05$; Fig. 1.4b). We also tested proinflammatory agent-elicited leukocyte migration in the Ndst1 mutant mice using an air-pouch model in which artificial chambers were created under the skin by injecting sterile air. In this model bacterial lipopolysaccharide (LPS, 1 mg/mouse)-induced neutrophil migration in the Ndst1 mutant mice was reduced to about half of the similarly treated wildtype control mice ($3.6 \times 10^5 \pm 1.1 \times 10^5$ versus $7.6 \times 10^5 \pm 1.9 \times 10^5$, respectively, $P < 0.05$). Taken together, all three mouse models tested consistently showed attenuated inflammatory responses in the Ndst1 mutant mice, thus demonstrating that heparan sulfate critically modulates inflammation in vivo.

Our characterization showed that TekCre-mediated Ndst1 ablation occurred in both endothelial cells and leukocytes (Wang et al. 2005; Fuster et al. 2007). To further differentiate the regulatory role of endothelial and leukocyte heparan sulfate in the inflammatory response, bone marrow transplantation experiments were performed by generating chimeric mice in which Ndst1 was selectively inactivated in either endothelial cells or leukocytes. To generate the true "endothelial-specific Ndst1" mutant, the lethally irradiated Ndst1 mutant mice were transplanted with

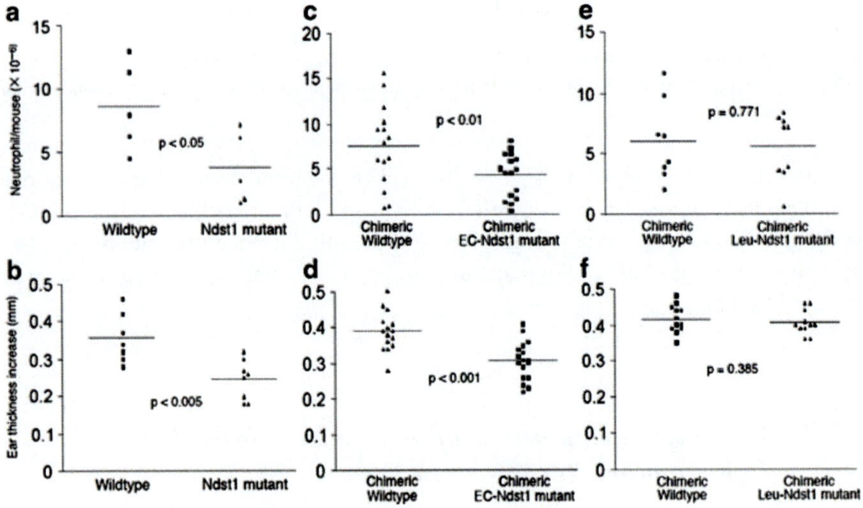

Fig. 1.4 Endothelial but not leukocyte Ndst1 deficiency impairs inflammatory response in vivo. (**a**, **c**, **e**) Thioglycollate-induced acute peritonitis. (**b**, **d**, **f**) Allergic contact dermatitis. Chimeric wildtype, chimeric mice with wildtype endothelial cells and leukocytes; Chimeric EC-Ndst1 mutant, chimeric mice with Ndst1 deficiency restricted in endothelial cells; Chimeric Leu-Ndst1 mutant, chimeric mice with Ndst1 deficiency restricted in leukocytes

wildtype bone marrow to restore heparan sulfate synthesis in leukocytes while retaining the Ndst1 deficiency in endothelial cells. Similarly, the irradiated wildtype mice were transplanted with Ndst1 mutant bone marrow to produce true "leukocyte-specific Ndst1 mutant" mice. A control wildtype group was also generated by transplanting wildtype bone marrow to irradiated wildtype recipient mice. At 6–8 weeks after the transplantation, when the peripheral hematology parameters appeared normal, the chimeric mice were examined for inflammatory responses using thioglycollate-induced acute peritonitis and allergic contact dermatitis models. We observed that the true "endothelial-specific Ndst1" mutant mice fully recapitulated the impaired inflammatory responses observed in the Ndst1 mutant mice (Fig. 1.4c, d), establishing that endothelial heparan sulfate critically modulates inflammatory response in vivo. Intriguingly, the true "leukocyte-specific Ndst1" mutant mice displayed a normal response (Fig. 1.4e, f), suggesting that leukocyte heparan sulfate might not be involved in the inflammatory response in vivo. However, this result cannot rule out the possibility that leukocyte heparan sulfate might also regulate inflammation (Tanaka et al. 1996) since Ndst1 ablation only partially alters heparan sulfate structure. To unequivocally resolve this issue, further studies that completely eliminate heparan sulfate structure on leukocytes, such as ablation of heparan sulfate biosynthesis copolymerase Ext1 or Ext2, will be required.

1.3.3 Endothelial Heparan Sulfate Interacts with L-Selectin to Mediate Leukocyte Rolling During Inflammation

Studies have established that the inflammatory response is a multistep process in which each step is mediated and coordinated by different groups of molecules (Ebnet et al. 1996; Butcher and Picker 1996; Parish 2006). To define the steps of the inflammatory response that were modulated by endothelial heparan sulfate, we carried out a parallel flow chamber assay in vitro that mimics the leukocyte–endothelial cell interaction during inflammation in vivo (Ebnet et al. 1996). Wildtype bone marrow neutrophils were perfused to flow through the tumor necrosis factor-α (TNFα) activated mouse lung endothelial cell monolayer. Under shear stresses of 1–2 dyne/cm^2, a small fraction of neutrophils were attached and rolled slowly along the endothelial cells and underwent multiple discrete interruptions in motion. The number of rolling cells did not differ on the Ndst1 mutant versus on the wildtype endothelial cells (Fig. 1.5a). However, the rolling velocity was about 40% greater on Ndst1 mutant endothelial cells at shear stress of 1 dyne/cm^2 (27.5 ± 3.1 μm/s versus 19.2 ± 2.3 μm/s; $P < 0.05$; Fig. 1.5a). When the sulfation of heparan sulfate was completely abrogated in endothelial cells by inactivation of both Ndst1 and Ndst2, the rolling velocity at 1 dyne/cm^2 was increased to 60% faster (38.6 ± 0.3 μm/s versus 24.0 ± 0.8 μm/s; Fig. 1.5c). Interestingly, this difference in velocity vanished when the flow rate was increased to shear stress at

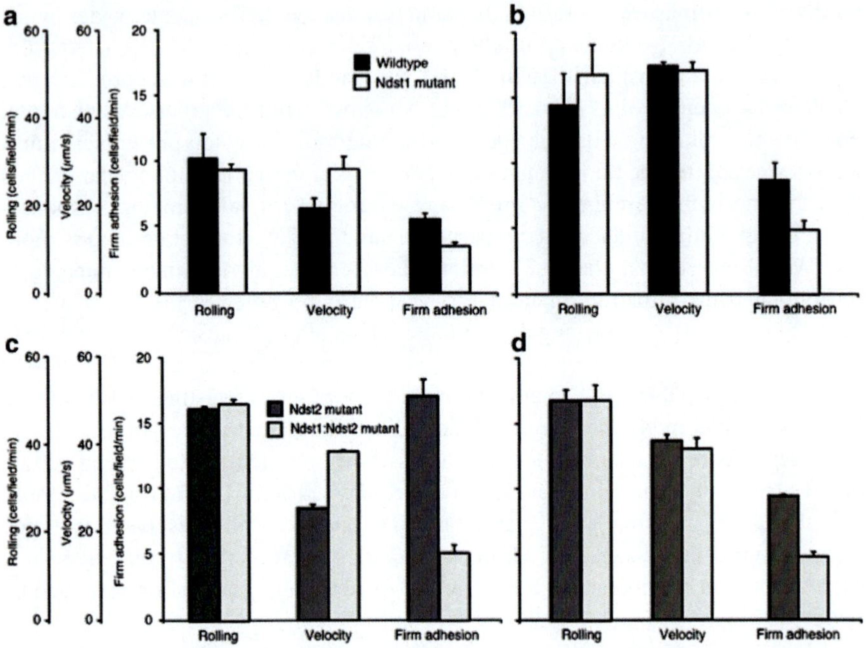

Fig. 1.5 Endothelial Ndst1 deficiency alters cell surface neutrophil rolling and firm adhesion under flow. Mouse bone marrow neutrophils from wildtype mice (**a–c**) or L-selectin-deficient mice (**d**) were flowed across TNFα-activated endothelial cells at shear stress of 1 dyne/cm^2 (**a**, **c**, **d**) or 2 dyne/cm^2 (**b**). Data were collected from three independent experiments and are presented as mean ± SE

2 dyne/cm^2 (51.9 ± 0.9 μm/s versus 50.9 ± 1.9 μm/s; $P = 0.88$; Fig. 1.5b), suggesting that heparan sulfate may only modulate leukocyte rolling at the vascular bed with a very low flow rate, such as at the location of postcapillary venules where leukocytes migrate out into perivasculature during inflammation.

L-selectin is constitutively expressed on neutrophil cell surfaces and participates in initial leukocyte rolling during inflammation. Studies have shown that recombinant L-selectin interacts in vitro with heparin and heparan sulfate and the interaction depends on the sulfation modification (Nelson et al. 1993; Borsig et al. 2002; Koenig et al. 1998; Giuffre et al. 1997; Norgard-Sumnicht and Varki 1995; Norgard-Sumnicht et al. 1993), suggesting that endothelial heparan sulfate may interact with L-selectin on leukocytes to modulate the L-selectin-mediated neutrophil rolling step in inflammation. This was tested directly. Indeed, direct binding of recombinant L-selectin to Ndst1 mutant endothelial cells was reduced by 27.8 ± 1.2% compared to that of wildtype control endothelial cells (Fig. 1.6). This was further confirmed by examining endothelial cells that are deficient in both Ndst1 and Ndst2. Deficiency in Ndst2 alone had no effect on the binding of L-selectin. However, the deficiency of both Ndst1 and Ndst2 resulted in a further reduction of L-selectin binding compared to Ndst1 mutant endothelial cells, reaching

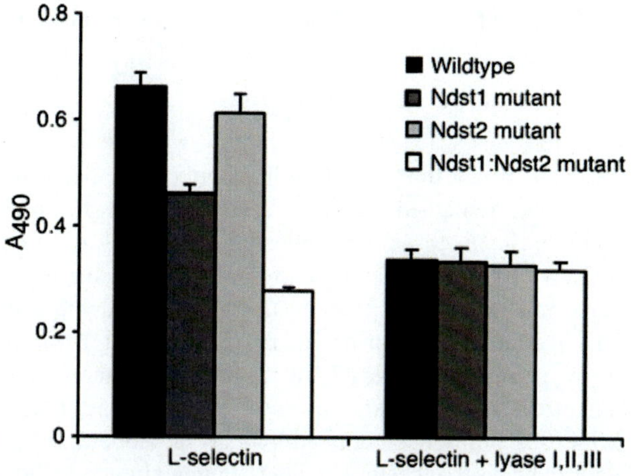

Fig. 1.6 Endothelial Ndst1 deficiency reduces cell surface heparan sulfate-dependent binding of L-selectin. TNFα-activated confluent mouse endothelial cells were incubated with recombinant human L-selectin protein, followed by HRP-conjugated anti-L-selectin antibody. The cells treated with heparin lyases I–III that degraded cell surface heparan sulfate served as background control

57.8 ± 1.2%, which was comparable to the cells treated with heparin lyases (Fig. 1.6). To alternatively confirm that the alteration in the L-selectin–endothelial heparan sulfate interaction accounted for the rolling velocity phenotype, The flow assay was repeated with L-selectin deficient neutrophils at 1 dyne/cm^2. L-selectin deficient neutrophils showed a more enhanced rolling velocity than wildtype neutrophils on activated Ndst2 deficient endothelial cells which possessed a normal heparan sulfate structure (40.6 ± 1.3 μm/s versus 24.0 ± 0.8 μm/s; $P < 0.05$; Fig. 1.5c). However, the rolling velocity difference of wildtype neutrophils on Ndst2-deficient and Ndst1:Ndst2 double-deficient endothelial cells vanished when L-selectin deficient neutrophils were perfused (38.6 ± 2.6 μm/s versus 40.6 ± 1.3 μm/s, respectively; $P = 0.61$. Fig. 1.5d), further indicating that L-selectin-mediated leukocyte rolling is modulated by endothelial heparan sulfate. Together, these observations directly showed that L-selectin interacts with endothelial heparan sulfate to modulate neutrophil rolling. Interestingly, although L-selectin has been clearly shown to mediate the initial leukocyte tethering and rolling during inflammation, no natural ligand for L-selectin has been identified at the postcapillary venules (Varki 1994; Hemmerich et al. 2001). Therefore, our observations demonstrate that endothelial heparan sulfate functions as a natural L-selectin ligand during inflammation, suggesting that the interaction with L-selectin may represent a major molecular mechanism underlying the regulatory function of endothelial heparan sulfate on initial leukocyte rolling during inflammation. Collectively, these serial experiments establish that endothelial heparan sulfate mediates the initial leukocyte rolling in inflammation by interacting with L-selectin.

1.3.4 Endothelial Heparan Sulfate Presents Chemokines on Luminal Surfaces to Mediate Leukocyte Firm Adhesion During Inflammation

In our flow assay, a small fraction of rolling leukocytes became permanently adherent. The ones that remained stationary at a given point for extended periods of time (>30 s) were counted as firm adhesion. According to this criterion, we consistently noticed ~40% reduction of neutrophils that became firmly adhered on the Ndst1 mutant endothelial cells than on the control wildtype endothelial cells (Fig. 1.5a). This reduction was further increased to about 70% when wildtype neutrophils were perfused on Ndst1:Ndst2 double mutant endothelial cells (Fig. 1.5c). This effect was not altered by shear stress changes (1–2 dyne/cm^2) or by perfusing with L-selectin-deficient neutrophils (Fig. 1.5b, d), indicating that the alteration of neutrophil firm adhesion on Ndst1-deficient endothelial cells was not caused by the impaired interaction between L-selectin and Ndst1 mutant endothelial heparan sulfate. The firm adhesion of leukocytes on inflamed endothelial cells has been known to be mainly caused by chemokine-elicited activation of G-protein-coupled receptors and integrins expressed on leukocyte cell membranes. This allows rolling leukocytes to interact with adhesion molecules, including intercellular adhesion molecules-1 and -2 (ICAM-1, ICAM-2) and vascular cell adhesion molecule 1 (VCAM-1), to induce cell spreading and firm adhesion (Ebnet et al. 1996; Rot and von Andrian 2004). Examination of ICAM-1, ICAM-2, and VCAM-1 expression on the endothelial cell surface using antibody staining did not reveal any obvious difference between the Ndst1 mutant and the wildtype endothelial cells. Chemokines KC and MIP2 are known to be implicated in neutrophil recruitment during inflammation in mice (Remick et al. 2001). Stimulation of the Ndst1 mutant and wildtype endothelial cells with TNF-α resulted in the same amount of KC and MIP2 release. Thus, these observations ruled out the possibility that Ndst1 ablation might alter expression of adhesion molecules and chemokine production to cause the attenuated neutrophil firm adhesion.

Chemokines have been known to bind heparin and heparan sulfate (Rot and von Andrian 2004), therefore we reasoned that the reduction in firm adhesion of neutrophils might be caused by the altered cell surface chemokine presentation on Ndst1 mutant endothelial cells. This was tested by measuring the direct bindings of the mouse chemokine KC and human chemokine interleukin 8 (IL-8) to the endothelial cell surface. Both KC and IL-8 showed about 50% reduction in their binding on the Ndst1 mutant compared to wildtype control endothelial cells (Fig. 1.7). Parallel experiments employing heparin as a competitive inhibitor of cell surface heparan sulfate or treating the cells with heparinase prior to the binding assay consistently demonstrated that the reduced cell surface binding of chemokine depended completely on heparan sulfate, showing that Ndst1 ablation disrupted cell surface chemokine binding and presentation. Consistent with these in vitro observations, Ndst1 mutant mice displayed disrupted neutrophil recruitment in the air-pouch model in which mouse chemokine KC or MIP-2 was injected to induce neutrophil

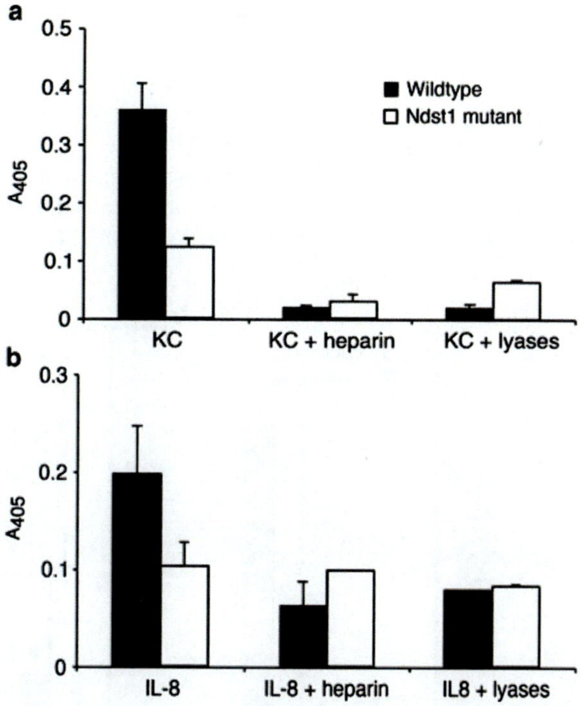

Fig. 1.7 Endothelial Ndst1 deficiency attenuates cell surface chemokine immobilization and presentation. TNFα-activated confluent mouse endothelial cells were incubated with chemokine KC (**a**) or IL-8 (**b**), and followed by the HRP-conjugated specific antibody to the respected chemokine. The cells treated with heparin lyases I–III that degraded cell surface heparan sulfate or high concentrations of heparin served as background control

infiltration. Taken together, these serial studies demonstrate that Ndst1 ablation reduced cell surface chemokine presentation, which, as a consequence, impaired chemokine-mediated neutrophil firm adhesion, thus revealing a second molecular mechanism for endothelial heparan sulfate to modulate inflammation in vivo.

1.3.5 Endothelial Heparan Sulfate Participates in Abluminal-to-Luminal Transcytosis of Chemokine During Inflammation

During inflammation chemokines released from resident cells, such as fibroblasts, smooth muscle cells and epithelial cells, serve as the major chemokine source for leukocyte recruitment (Garcia-Ramallo et al. 2002; Buckley et al. 2001). This indicates that the released chemokines need to diffuse through the extracellular matrix, reach the abluminal side of endothelial cells, be transported from the abluminal side through endothelial cell layers, and then present on luminal cell surfaces to mediate leukocyte firm adhesion. Previous studies using electron microscopy suggested that endothelial heparan sulfate might be involved in the directional transcytosis of injected IL-8 across endothelial cells (Middleton et al. 1997). To directly determine whether endothelial heparan sulfate participates in chemokine transcytosis, we employed the Boyden Chamber assay with confluent endothelial cells cultured on membranes to examine if Ndst1 ablation affects IL-8 transcytosis in vitro. [^{125}I]IL-8 was supplemented to the lower compartment and the radioactivity that appeared in the upper compartment was quantified. At 37°C for 30 min, ~25% less [^{125}I]IL-8 was transported to the top compartment in chambers containing Ndst1 mutant endothelial cells than in chambers containing wildtype control cells (Fig. 1.8a; $P < 0.05$). Similar results were obtained when the experiments were carried out at 37°C for 15 min. In the Boyden Chamber assay, transportation of

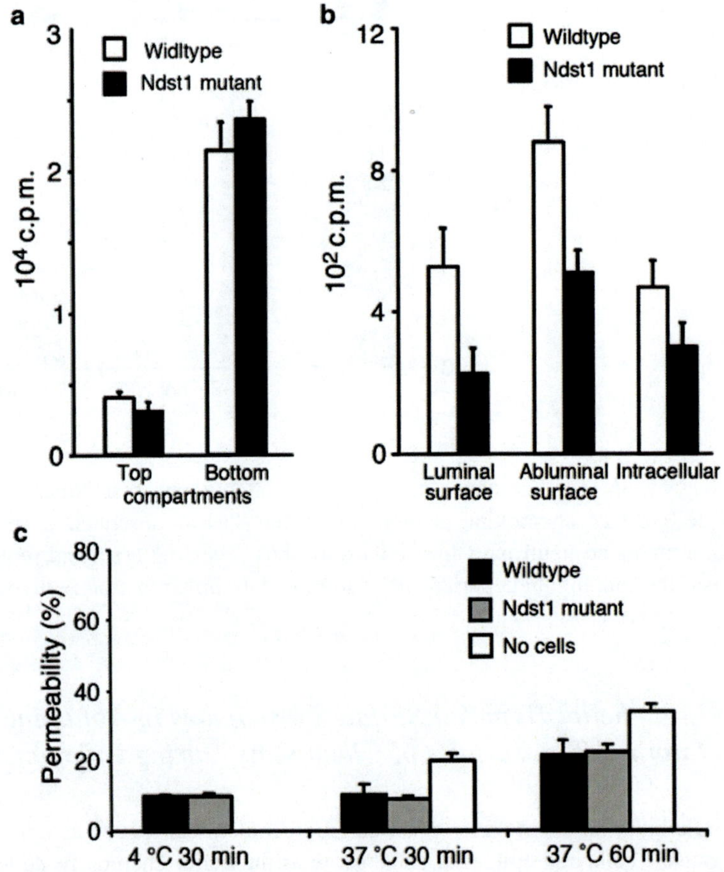

Fig. 1.8 Endothelial Ndst1 deficiency alters transcytosis but not pericellular transport of chemokine. (**a**) Distribution of [^{125}I]IL-8 in the top and bottom chambers after the Boyden Chamber assay was carried out for 30 min at 37°C. There was a small decrease in the chemokine transported (via both transcytosis and pericellular path) across Ndst1 mutant endothelial cells compared to the wildtype control. (**b**) IL-8 bound on the Ndst1 mutant cell surfaces and in the intracellular pool was reduced about 50% compared to the wildtype control. (**c**) Endothelial Ndst1 deficiency did not affect endothelial cell permeability. The experiment described in (**a**) was carried out at both 4°C (a condition to block vesicle-mediated transcellular transport) and 37°C with dextran sulfate-FITC. No difference was observed between the Ndst1 mutant and wildtype endothelial cells

chemokines from the lower compartment to the upper compartment can be routed by diffusion through intercellular gaps and/or by vesicle-mediated transcytosis paths. To differentiate the contribution of the two potential routes to the observed reduction in chemokine transportation, we initially determined whether Ndst1 deficiency alters endothelial cell permeability, a measurement of pericellular diffusion. No difference in cell permeability was observed between the Ndst1 mutant and wildtype endothelial cells. This showed that the pericellular path of chemokine transportation was not affected by Ndst1 ablation. This observation also suggested that the reduced chemokine transportation into the top compartment by the Ndst1 mutant endothelium is caused by disruption of chemokine transcytosis. This was further tested by measuring the IL-8 that bound to the bottom and top surfaces of cells, and in the cytoplasm. Compared to the wildtype control, the amount of IL-8 that bound to both the top and bottom surfaces as well as to the intracellular portion of the Ndst1 mutant endothelial cells was decreased by about 50%. This shows that heparan sulfate participates in chemokine transcytosis by mediating internalization and abluminal-to-luminal transport across endothelial cells, thus revealing the third mechanism for endothelial heparan sulfate in the regulation of inflammation.

1.3.6 Endothelial Heparan Sulfate Directs Intraluminal Crawling of Neutrophils by Retaining a Chemotactic Chemokine Gradient

An additional step in the leukocyte recruitment cascade was recently observed that bridged adhesion and diapedesis, namely Mac-1-mediated intraluminal crawling. In this process, neutrophils crawl on the endothelium prior to transmigration through endothelial junctions. If crawling was disabled, diapedesis was delayed. Very recently, Massena et al. (2010) examined whether endothelial heparan sulfate plays any role in this intraluminal neutrophil crawling and transmigration by introducing a chemokine (MIP-2) gradient on the cremeaster muscle of anesthetized wildtype mice or heparanase overexpressing mice which produce truncated heparan sulfate side chains (Massena et al. 2010). Neutrophil–endothelial interactions were visualized by intravital microscopy and chemokine gradients detected by confocal microscopy. It was observed that the release of the localized extravascular chemokine induced directed neutrophil crawling along a chemotactic gradient immobilized on the endothelium and accelerated their recruitment into the target tissue compared to the homogenous extravascular chemokine concentration (MIP-2 superfusion). More interestingly, endothelial chemokine sequestration occurred exclusively in venules and was heparan sulfate-dependent, and neutrophils in heparanase overexpressed mice exhibited random crawling. The latter observations clearly showed that endothelial heparan sulfate sequesters chemokine and forms a chemokine gradient on luminal surfaces, which in turn, directs crawling leukocytes toward transmigration loci (Massena et al. 2010). Therefore, these observations reveal that endothelial heparan sulfate also functions to form chemokine gradients on luminal cell surfaces to direct intraluminal crawling of leukocytes during inflammation.

In summary, employing the mouse genetic approach, we and others demonstrated that endothelial heparan sulfate critically modulates leukocyte trafficking during inflammation in vivo. Mechanistic studies further revealed that heparan sulfate functions as a multifaceted modulator in inflammation, including acting as a ligand for L-selectin to mediate neutrophil rolling, facilitating abluminal-to-luminal chemokine transcytosis, and binding, presentation, and formation of chemokine gradients on luminal surfaces to direct intraluminal crawling and to induce firm adhesion.

1.4 The Role of Heparan Sulfate in Angiogenesis

Angiogenesis refers to the generation of new blood vessels from existing ones, a process that differs from vasculogenesis (the de novo formation of blood vessels from mesoderm or endothelial cell progenitors) (Carmeliet 2005; Coultas et al. 2005; Risau 1997). Physiological angiogenesis occurs mainly during embryonic development and in the postnatal growth of tissues such as retina (Drake 2003; Carmeliet 2005). In adults, it takes place primarily in the female reproductive system and in the intestinal villa (Stappenbeck et al. 2002; Carmeliet et al. 1996). Abnormal angiogenesis has been observed to play critical roles in the pathogenesis of many human diseases, such as cancer, ischemic vascular disorders, psoriasis, arthritis and diabetic retinopathy (Folkman 1995; Carmeliet 2005). Elucidation of the mechanisms that control angiogenesis provides a basis for the development of novel treatments for angiogenesis-based human diseases.

1.4.1 Deficiency of Endothelial Ndst1 Does Not Affect Physiological Angiogenesis In Vivo

Biochemical studies using purified heparan sulfate and cells have shown in vitro that heparan sulfate interacts with major angiogenic molecules, such as VEGF, to critically regulate angiogenesis. These observations have been supported by alternative studies showing that mice lacking heparan sulfate-binding splice forms of VEGF (the VEGF120 mice) exhibit developmental angiogenesis defects (Carmeliet et al. 1999; Gerhardt et al. 2003), and that mouse embryonic stem cells deficient in Ndst1 and Ndst2 cannot form neovasculature in embryoid bodies (Jakobsson et al. 2006). However, the precise role of heparan sulfate in angiogenesis in vivo has been largely unknown. Our Ndst1 mutant mice bear heparan sulfate alteration only in the endothelium and leukocytes, thus providing an ideal system to determine the role of heparan sulfate in angiogenesis. Interestingly, the Ndst1 mutant mice appear normal with unaltered fertility, showing that the Ndst1 deficiency is compatible with developmental and reproductive angiogenesis. To determine whether the Ndst1 deficiency affects physiological angiogenesis in adult mice, the mice were tested with the cutaneous wound healing model. There was no observable difference between the Ndst1 mutant and the wildtype control mice when measuring the wound closure rate, clustering of PECAM-1 positive cells and the vascular density during resolution of the wounds. Together, these observations consistently show that Ndst1 deficiency does not affect physiological angiogenesis in vivo.

1.4.2 Deficiency of Endothelial Ndst1 Disturbs Tumor Angiogenesis In Vivo

Solid tumors and their metastasis require the process of angiogenesis, termed specifically as tumor angiogenesis, to invest themselves with host microvasculature and to support tumor growth (Carmeliet 2005). To determine whether endothelial heparan sulfate affects tumor angiogenesis, syngeneic Lewis lung carcinoma (LLC) cells were injected subcutaneously in the hindquarters of Ndst1 mutant and littermate wildtype control mice. LLC tumor growth in Ndst1 mutant mice was significantly attenuated compared to that grown in wildtype control mice (Fig. 1.9a). The effect was similarly seen when syngeneic mouse melanoma (B16BL6) tumor cells were injected. In the Ndst1 mutant mice, Ndst1 deficiency occurred in both endothelial cells and leukocytes. To differentiate the contribution of endothelial and leukocyte heparan sulfate to the attenuated tumor growth phenotype, the experiments were repeated with bone marrow transplanted chimera mice in which the Ndst1 deficiency was restricted to endothelial cells or leukocytes. The attenuated tumor growth occurred in the chimeric endothelial- but not in the leukocyte-specific Ndst1-deficient mice (Fig. 1.9b–d), indicating that endothelial, but not the leukocyte heparan sulfates, facilitates tumor growth. The implication of endothelial heparan sulfate in tumor growth suggests that endothelial heparan sulfate may modulate tumor angiogenesis to achieve its regulatory role in tumor growth. Therefore, the vascular pattern and density in tumor tissues were examined by immunostaining for PECAM-1, a specific marker for endothelial cells. Microvascular density was significantly reduced in tumors grown in the Ndst1 mutant mice compared to the ones grown in the wildtype control mice (Fig. 1.10a). This observation was further confirmed by examining the Ndst1 mutant mice with LLC tumor spheroids implanted in dorsal skinfold chambers, a model by which in vivo tumor angiogenesis can be kinetically monitored (Fig. 1.10b). Collectively, these in vivo observations clearly established that endothelial heparan sulfate functions to facilitate tumor growth by enhancing tumor angiogenesis in vivo.

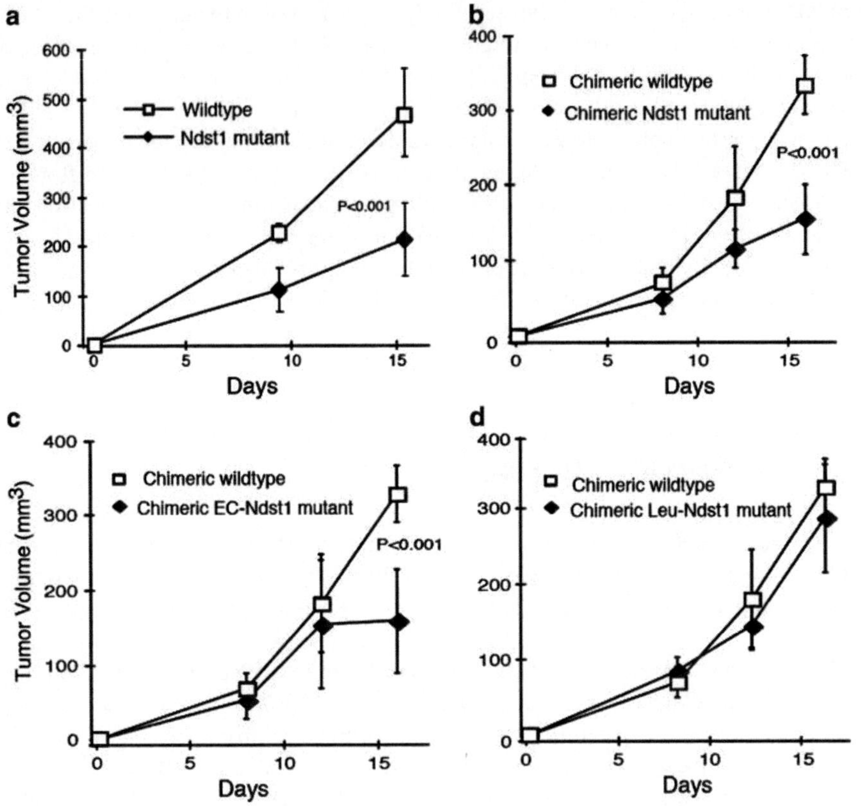

Fig. 1.9 Tumor growth is attenuated in Ndst1 mutant mice. (**a**) LLC tumor growth in Ndst1 mutant and wildtype control mice. Tumor volume was measured over days after LLC cell injection. (**b–d**) Tumor growth in bone marrow transplanted chimeric mice. Chimeric wildtype, both endothelial cells and leukocytes were wildtype; Chimeric EC-Ndst1 mutant, chimeric mice with Ndst1 deficiency restricted in endothelial cells; Chimeric Leu-Ndst1 mutant, chimeric mice with Ndst1 deficiency restricted in leukocytes

1.4.3 Deficiency of Endothelial Ndst1 Disturbs VEGF Signaling Which Correlates with Enhanced Endothelial Apoptosis in Tumor Vasculature

Many proangiogenic growth factors, in particular the heparan sulfate-binding splice isoforms of VEGF (such as VEGF165) bind to HSPGs expressed on tumor endothelial cells and in the extracellular matrix. This suggests that endothelial heparan sulfate might facilitate the formation of VEGF–VEGF receptor signaling complexes, thereby enhancing tumor angiogenesis in vivo. To test this hypothesis, the tumor tissues were initially examined to determine whether the Ndst1 ablation altered the heparan sulfate structure in tumor vasculature. Staining tumor tissues

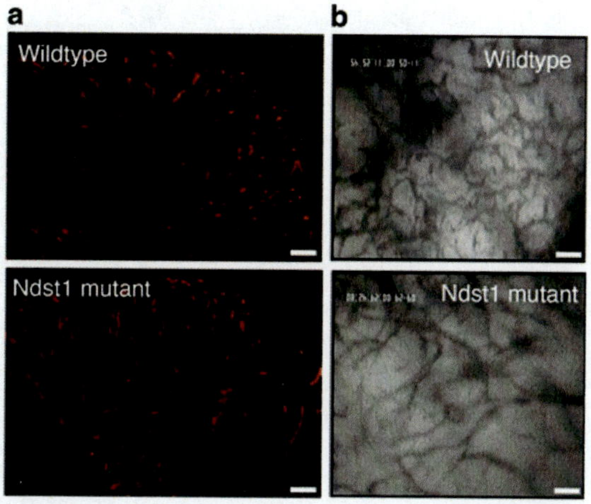

Fig. 1.10 Endothelial Ndst1 deficiency attenuates tumor angiogenesis. (**a**) Immunofluorescence staining of LLC tumor tissues with anti-PECAM-1 antibody to examine vasculature. (**b**) Intravital videomicroscopy of tumor vasculature formed in the skinfold chamber model 12 days after implantation with LLC tumor cell spheroids

with FGF-2 showed that FGF-2 binding to Ndst1-deficient tumor vessels was significantly reduced (Fig. 1.11a), which was consistent with FGF-2 staining of isolated Ndst1-deficient endothelial cells in vitro, and that the Ndst1 deficiency altered heparan sulfate structure in tumor vasculature. Since the binding of FGF-2 to heparan sulfate essentially depends on *N*- and 2-*O*-sulfates (Ishihara et al. 1992; Bai and Esko 1996), this observation also implied that Ndst1 ablation reduced the *N*- and/or 2-*O*-sulfation of tumor endothelial heparan sulfate. To determine whether the altered tumor endothelial heparan sulfate affects VEGF signaling locally, the tumor tissues were immunostained with the GV39M monoclonal antibody, which binds to the VEGF/VEGF receptor 2 complexes (Brekken et al. 1998; Joyce et al. 2005). A subset of wildtype tumor endothelial cells stained strongly for GV39M, whereas this was rarely seen in Ndst1 mutant tumor endothelial cells (Fig. 1.11b), indicating that Ndst1 ablation disrupted VEGF signaling during tumor angiogenesis in Ndst1 mutant mice. This in vivo observation was also supported by staining isolated endothelial cells with VEGF164 in which VEGF164 binding to the Ndst1 mutant endothelial cell surface was significantly reduced compared to the wildtype control.

VEGF signaling is critically required for endothelial cell survival (Coultas et al. 2005). Therefore, endothelial apoptosis in tumor tissues was also examined. In agreement with the disturbed VEGF signaling, a portion of the Ndst1-deficient tumor endothelial cells went apoptosis (Fig. 1.11c). However, this was not seen in wildtype tumor endothelial cells, revealing that Ndst1 ablation disrupts cell surface VEGF signaling which, as a consequence, causes tumor endothelial cell death. Taken together, these observations delineated the critical role of endothelial heparan sulfate in tumor angiogenesis, and also raises the rationale that selective blockage of endothelial heparan sulfate may be an effective approach to treat tumor angiogenesis-based human cancers.

In summary, these serial studies demonstrated that Ndst1 deficiency disrupted VEGF signaling, which, as a consequence, led to endothelial apoptosis, attenuation

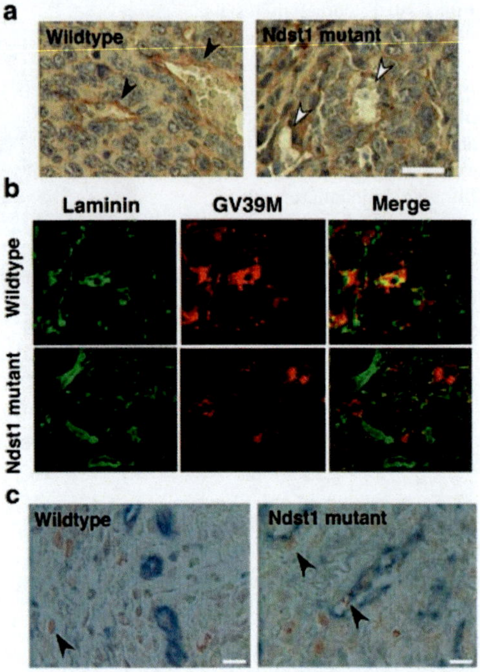

Fig. 1.11 Ndst1 deficiency alters heparan sulfate structure in tumor vasculature leading to impaired VEGF signaling and enhanced apoptosis of tumor endothelial cells. (**a**) In situ FGF-2 staining. Arrowheads indicate FGF-2 binding to vasculature. (**b**) Immunofluorescence staining for laminin, a component of vascular basement membranes, and for local VEGF signaling by GV39M, a monoclonal antibody recognizing VEGF–VEGF receptor complexes. (**c**) PECAM-1 (*blue*) and TUNEL (*brown*) staining to assess apoptotic endothelial cells in tumor tissues

of tumor angiogenesis and the final outcome of impaired tumor growth, thus establishing that endothelial heparan sulfate functions to promote tumor angiogenesis in vivo. Interestingly, the ablation of Ndst1 in mouse endothelial cells did not affect physiological angiogenesis and displayed a selective inhibitory effect on tumor angiogenesis, suggesting that targeting endothelial heparan sulfate may represent a novel approach in treating cancer with least effect on normal tissues.

1.5 Conclusions and Prospective

Our studies employing heparin and heparinoids revealed that non-anticoagulant heparins efficiently inhibit both P- and L-selectin-mediated leukocyte trafficking in inflammation, suggesting that non-anticoagulant heparins can be developed as potent anti-inflammatory drugs. This prospective has been supported by studies conducted by others (Zhou et al. 2002; Yuan et al. 2003; Frank et al. 2006; Ludwig et al. 2006b; Mocco et al. 2007; Young 2008; Brito et al. 2008). The non-anticoagulant heparin may be applied to block other P- and L-selectin-mediated pathological processes, such as tumor metastasis. Recent studies conducted by several groups consistently showed that non-coagulant heparins potently inhibited tumor metastasis in vitro and in vivo (Wei et al. 2004, 2005; Gao et al. 2006; Kragh and Loechel

2005; Ludwig et al. 2006a; Borsig 2007; Hostettler et al. 2007; Lee et al. 2008; Stevenson et al. 2007). Since heparin has been safely and widely used clinically for over 70 years, these studies highlight the bright future for the development of heparinoids as novel anti-inflammation and antitumor metastasis drugs. However, additional studies are required to define the fine structure of heparinoids that interacts with P- and L-selectin, and to develop more structurally defined heparinoids for clinical treatment.

By selective ablation of Ndst1 in mouse endothelial cells and leukocytes we demonstrate that endothelial heparan sulfate, but not leukocyte heparan sulfate, plays critical roles in inflammation and tumor angiogenesis. The participation of endothelial heparan sulfate, but not leukocyte heparan sulfate, in inflammation was further confirmed by examining Ndst1 mutant mice with an aortic allograft transplant rejection model (Dai et al. 2010). However, the non-involvement of leukocyte heparan sulfate in inflammation and in physiological angiogenesis needs to be carefully interpreted since Ndst1 ablation only partially reduces sulfation modification (Rollins 1997). Complete ablation of heparan sulfate (such as knockout of Ext1 or Ext2) is required to unequivocally determine the role of heparan sulfate in these biological processes.

In our studies, we observed that endothelial heparan sulfate functions as a ligand when participating in L-selectin-mediated leukocyte rolling. However, heparinase treatment of endothelial cells only partially reduced cell surface L-selectin binding (Fig. 1.6), indicating that endothelial cells also express non-heparan sulfate structure ligand for L-selectin. This observation suggests that endothelial heparan sulfate may cowork with other unidentified L-selectin ligands to modulate L-selectin functions during inflammation. This prospection has been supported by our other findings showing that deficiency of endothelial Ndst1 or Ndst1 and Ndst2 disrupted L-selectin-mediated leukocyte rolling only at a shear stress of 1 dyne/cm^2 (Fig. 1.5). Further studies to identify the unknown L-selectin ligand expressed by endothelial cells and to examine the functional relationship of the unknown L-selectin ligand with endothelial heparan sulfate during inflammation will lead to further understanding of the molecular mechanism of L-selectin-mediated leukocyte trafficking.

Our studies also established that endothelial heparan sulfate functions to transcytose and to present chemokines during inflammation. It is well known that leukocyte subsets are selectively recruited into inflammatory sites or immune organs such as spleen, lymph node and thymus by specific chemokines (Springer 1994; Butcher and Picker 1996; Ebnet et al. 1996). The structure of heparan sulfate is known to be tissue/organ- and developmental stage-specific (Allen et al. 2001; Ledin et al. 2004). Therefore, it would be important to determine whether heparan sulfate participates in leukocyte subtype entry into different tissues.

Instead of altering heparan sulfate structure as we did in our studies, other groups examined the roles of proteoglycans in inflammation and angiogenesis by ablation or overexpression of individual proteoglycans, such as syndecan-1 (Li et al. 2002; Gotte et al. 2002; Park et al. 2000a; Kharabi Masouleh et al. 2009; Elenius et al. 2004), syndecan-4 (Ishiguro et al. 2000a; Echtermeyer et al. 2001), perlecan (Zhou et al. 2004), or collagen XVIII (Fukai et al. 2002). In general, the studies of syndecan-4, perlecan and collagen XVIII mutant mice are in good agreement with our Ndst1 studies. For example, syndecan-4 knockout mice display impaired fetal angiogenesis in the placenta labyrinth (Ishiguro et al. 2000b) and during wound healing (Ishiguro et al. 2000a; Echtermeyer et al. 2001). In contrast, the syndecan-1 mutant mice have shown the phenotypes that are opposite to what we have observed in our Ndst1 mutant mice. Syndecan-1 deficiency in mice enhances inflammatory responses (Gotte et al. 2002; Kharabi Masouleh et al. 2009) and overexpression of syndecan-1 disrupts angiogenesis in wound healing (Elenius et al. 2004), showing that syndecan-1 functions as a negative regulator of inflammation and angiogenesis in vivo. These studies also further showed that syndecan-1 expressed on leukocytes plays a dominant role in the anti-inflammatory function (Kharabi Masouleh et al. 2009; Gotte et al. 2002). In contrast, our studies showed that endothelial, but not leukocyte heparan sulfate, plays a major role in regulating inflammation. The differences in phenotypes displayed in syndecan-1 and our Ndst1 mutant mice can be explained by the disparity between the alterations of HSPG structure in the two mutants. In our Ndst1 mutant, heparan sulfate structure was globally changed in all HSPGs expressed on endothelial cells and leukocytes, whereas in the syndecan-1 mutant mice, only syndecan-1 is ablated and the other HSPGs expressed on the cell surface of endothelial cells and leukocytes remain unchanged or are even possibly up-regulated due to compensation. In our ongoing studies, we have observed that endothelial cells express several syndecans and glypicans. Furthermore, the syndecan-1 core possesses heparan sulfate-independent functions that are involved in cell adhesion (McQuade et al. 2006; Beauvais and Rapraeger 2003). Therefore, the syndecan-1 phenotype can be both heparan sulfate-dependent and -independent. In addition, syndecan-1 may also carry chondroitin sulfate chains (Bernfield et al. 1999). Thus, syndecan-1 deficiency may also be complicit in the disruption of chondroitin sulfate-mediated biological functions. Further studies, such as examining heparan sulfate/chondroitin sulfate-free syndecan-1 mutant mice, may help to further address this issue and to accommodate the conflicting observations between syndecan-1 and our Ndst1 mutant mice.

Acknowledgments I thank Ms. Karen Howard for her careful reading and English version of this manuscript. Our research is supported by the Georgia Cancer Coalition, the American Heart Association, and the NIH (R01HL093339 and P41RR005351).

References

Aikawa J, Grobe K, Tsujimoto M, Esko JD (2001) Multiple isozymes of heparan sulfate/heparin glcnac n-deacetylase/glcn n-sulfotransferase. Structure and activity of the fourth member, ndst4. J Biol Chem 276(8):5876–5882

Allen BL, Filla MS, Rapraeger AC (2001) Role of heparan sulfate as a tissue-specific regulator of fgf-4 and fgf receptor recognition. J Cell Biol 155(5):845–858

Bai X, Esko JD (1996) An animal cell mutant defective in heparan sulfate hexuronic acid 2-o-sulfation. J Biol Chem 271(30):17711–17717

Beauvais DM, Rapraeger AC (2003) Syndecan-1-mediated cell spreading requires signaling by alphavbeta3 integrins in human breast carcinoma cells. Exp Cell Res 286(2):219–232. doi: S0014482703001265 [pii]

Beauvais DM, Rapraeger AC (2004) Syndecans in tumor cell adhesion and signaling. Reprod Biol Endocrinol 2:3

Bernfield M, Gotte M, Park PW, Reizes O, Fitzgerald ML, Lincecum J, Zako M (1999) Functions of cell surface heparan sulfate proteoglycans. Annu Rev Biochem 68:729–777. doi:10.1146/annurev.biochem.68.1.729

Bishop JR, Schuksz M, Esko JD (2007) Heparan sulphate proteoglycans fine-tune mammalian physiology. Nature 446(7139):1030–1037

Borsig L (2007) Antimetastatic activities of modified heparins: selectin inhibition by heparin attenuates metastasis. Semin Thromb Hemost 33(5):540–546. doi:10.1055/s-2007-982086

Borsig L, Wong R, Feramisco J, Nadeau DR, Varki NM, Varki A (2001) Heparin and cancer revisited: mechanistic connections involving platelets, p-selectin, carcinoma mucins, and tumor metastasis. Proc Natl Acad Sci USA 98(6):3352–3357

Borsig L, Wong R, Hynes RO, Varki NM, Varki A (2002) Synergistic effects of l- and p-selectin in facilitating tumor metastasis can involve non-mucin ligands and implicate leukocytes as enhancers of metastasis. Proc Natl Acad Sci USA 99(4):2193–2198

Brekken RA, Huang X, King SW, Thorpe PE (1998) Vascular endothelial growth factor as a marker of tumor endothelium. Cancer Res 58(9):1952–1959

Brito AS, Arimateia DS, Souza LR, Lima MA, Santos VO, Medeiros VP, Ferreira PA, Silva RA, Ferreira CV, Justo GZ, Leite EL, Andrade GP, Oliveira FW, Nader HB, Chavante SF (2008) Anti-inflammatory properties of a heparin-like glycosaminoglycan with reduced anti-coagulant activity isolated from a marine shrimp. Bioorg Med Chem 16(21):9588–9595. doi:S0968-0896(08)00857-2 [pii] 10.1016/j.bmc.2008.09.020

Buckley CD, Pilling D, Lord JM, Akbar AN, Scheel-Toellner D, Salmon M (2001) Fibroblasts regulate the switch from acute resolving to chronic persistent inflammation. Trends Immunol 22(4):199–204. doi:S1471-4906(01)01863-4 [pii]

Butcher EC, Picker LJ (1996) Lymphocyte homing and homeostasis. Science 272(5258):60–66

Carmeliet P (2005) Angiogenesis in life, disease and medicine. Nature 438(7070):932–936

Carmeliet P, Ferreira V, Breier G, Pollefeyt S, Kieckens L, Gertsenstein M, Fahrig M, Vandenhoeck A, Harpal K, Eberhardt C, Declercq C, Pawling J, Moons L, Collen D, Risau W, Nagy A (1996) Abnormal blood vessel development and lethality in embryos lacking a single vegf allele. Nature 380(6573):435–439

Carmeliet P, Ng YS, Nuyens D, Theilmeier G, Brusselmans K, Cornelissen I, Ehler E, Kakkar VV, Stalmans I, Mattot V, Perriard JC, Dewerchin M, Flameng W, Nagy A, Lupu F, Moons L, Collen D, D'Amore PA, Shima DT (1999) Impaired myocardial angiogenesis and ischemic cardiomyopathy in mice lacking the vascular endothelial growth factor isoforms vegf164 and vegf188. Nat Med 5(5):495–502

Coultas L, Chawengsaksophak K, Rossant J (2005) Endothelial cells and vegf in vascular development. Nature 438(7070):937–945

Dai E, Liu LY, Wang H, McIvor D, Sun YM, Macaulay C, King E, Munuswamy-Ramanujam G, Bartee MY, Williams J, Davids J, Charo I, McFadden G, Esko JD, Lucas AR (2010) Inhibition of chemokine-glycosaminoglycan interactions in donor tissue reduces mouse allograft vasculopathy and transplant rejection. PLoS One 5(5):e10510. doi:10.1371/journal.pone.0010510

Drake CJ (2003) Embryonic and adult vasculogenesis. Birth Defects Res C Embryo Today 69 (1):73–82

Ebnet K, Kaldjian EP, Anderson AO, Shaw S (1996) Orchestrated information transfer underlying leukocyte endothelial interactions. Annu Rev Immunol 14:155–177

Echtermeyer F, Streit M, Wilcox-Adelman S, Saoncella S, Denhez F, Detmar M, Goetinck P (2001) Delayed wound repair and impaired angiogenesis in mice lacking syndecan-4. J Clin Invest 107(2):R9–R14. doi:10.1172/JCI10559

Elenius V, Gotte M, Reizes O, Elenius K, Bernfield M (2004) Inhibition by the soluble syndecan-1 ectodomains delays wound repair in mice overexpressing syndecan-1. J Biol Chem 279 (40):41928–41935. doi:10.1074/jbc.M404506200 M404506200 [pii]

Esko JD, Lindahl U (2001) Molecular diversity of heparan sulfate. J Clin Invest 108(2):169–173

Esko JD, Selleck SB (2002) Order out of chaos: assembly of ligand binding sites in heparan sulfate. Annu Rev Biochem 71:435–471

Folkman J (1995) Angiogenesis in cancer, vascular, rheumatoid and other disease. Nat Med 1 (1):27–31

Forsberg E, Kjellen L (2001) Heparan sulfate: lessons from knockout mice. J Clin Invest 108 (2):175–180

Forsberg E, Pejler G, Ringvall M, Lunderius C, Tomasini-Johansson B, Kusche-Gullberg M, Eriksson I, Ledin J, Hellman L, Kjellen L (1999) Abnormal mast cells in mice deficient in a heparin-synthesizing enzyme. Nature 400(6746):773–776

Frank RD, Holscher T, Schabbauer G, Tencati M, Pawlinski R, Weitz JI, Mackman N (2006) A non-anticoagulant synthetic pentasaccharide reduces inflammation in a murine model of kidney ischemia-reperfusion injury. Thromb Haemost 96(6):802–806. doi:06120802 [pii]

Fukai N, Eklund L, Marneros AG, Oh SP, Keene DR, Tamarkin L, Niemela M, Ilves M, Li E, Pihlajaniemi T, Olsen BR (2002) Lack of collagen xviii/endostatin results in eye abnormalities. EMBO J 21(7):1535–1544. doi:10.1093/emboj/21.7.1535

Fuster MM, Wang L, Castagnola J, Sikora L, Reddi K, Lee PH, Radek KA, Schuksz M, Bishop JR, Gallo RL, Sriramarao P, Esko JD (2007) Genetic alteration of endothelial heparan sulfate selectively inhibits tumor angiogenesis. J Cell Biol 177(3):539–549

Gallagher JT (2001) Heparan sulfate: growth control with a restricted sequence menu. J Clin Invest 108(3):357–361. doi:10.1172/JCI13713

Gao Y, Wei M, Zheng S, Ba X, Hao S, Zeng X (2006) Chemically modified heparin inhibits the in vitro adhesion of nonsmall cell lung cancer cells to p-selectin. J Cancer Res Clin Oncol 132 (4):257–264. doi:10.1007/s00432-005-0061-9

Garcia-Ramallo E, Marques T, Prats N, Beleta J, Kunkel SL, Godessart N (2002) Resident cell chemokine expression serves as the major mechanism for leukocyte recruitment during local inflammation. J Immunol 169(11):6467–6473

Gerhardt H, Golding M, Fruttiger M, Ruhrberg C, Lundkvist A, Abramsson A, Jeltsch M, Mitchell C, Alitalo K, Shima D, Betsholtz C (2003) Vegf guides angiogenic sprouting utilizing endothelial tip cell filopodia. J Cell Biol 161(6):1163–1177

Giuffre L, Cordey AS, Monai N, Tardy Y, Schapira M, Spertini O (1997) Monocyte adhesion to activated aortic endothelium: Role of l-selectin and heparan sulfate proteoglycans. J Cell Biol 136(4):945–956

Gotte M (2003) Syndecans in inflammation. FASEB J 17(6):575–591

Gotte M, Joussen AM, Klein C, Andre P, Wagner DD, Hinkes MT, Kirchhof B, Adamis AP, Bernfield M (2002) Role of syndecan-1 in leukocyte-endothelial interactions in the ocular vasculature. Invest Ophthalmol Vis Sci 43(4):1135–1141

Grobe K, Ledin J, Ringvall M, Holmborn K, Forsberg E, Esko JD, Kjellen L (2002) Heparan sulfate and development: differential roles of the n-acetylglucosamine n-deacetylase/n-sulfotransferase isozymes. Biochim Biophys Acta 1573(3):209–215

Hacker U, Nybakken K, Perrimon N (2005) Heparan sulphate proteoglycans: the sweet side of development. Nat Rev Mol Cell Biol 6(7):530–541

Hemmerich S, Bistrup A, Singer MS, van Zante A, Lee JK, Tsay D, Peters M, Carminati JL, Brennan TJ, Carver-Moore K, Leviten M, Fuentes ME, Ruddle NH, Rosen SD (2001) Sulfation of l-selectin ligands by an hev-restricted sulfotransferase regulates lymphocyte homing to lymph nodes. Immunity 15(2):237–247

Hoogewerf AJ, Kuschert GS, Proudfoot AE, Borlat F, Clark-Lewis I, Power CA, Wells TN (1997) Glycosaminoglycans mediate cell surface oligomerization of chemokines. Biochemistry 36 (44):13570–13578

Hostettler N, Naggi A, Torri G, Ishai-Michaeli R, Casu B, Vlodavsky I, Borsig L (2007) P-selectin- and heparanase-dependent antimetastatic activity of non-anticoagulant heparins. FASEB J 21(13):3562–3572. doi:fj.07-8450com [pii] 10.1096/fj.07-8450com

Iozzo RV, San Antonio JD (2001) Heparan sulfate proteoglycans: heavy hitters in the angiogenesis arena. J Clin Invest 108(3):349–355

Ishiguro K, Kadomatsu K, Kojima T, Muramatsu H, Nakamura E, Ito M, Nagasaka T, Kobayashi H, Kusugami K, Saito H, Muramatsu T (2000a) Syndecan-4 deficiency impairs the fetal vessels in the placental labyrinth. Dev Dyn 219(4):539–544. doi:10.1002/1097-0177(2000) 9999:9999 <::AID-DVDY1081>3.0.CO;2-K [pii] 10.1002/1097-0177(2000) 9999:9999<::AID-DVDY1081>3.0.CO;2-K

Ishiguro K, Kadomatsu K, Kojima T, Muramatsu H, Tsuzuki S, Nakamura E, Kusugami K, Saito H, Muramatsu T (2000b) Syndecan-4 deficiency impairs focal adhesion formation only under restricted conditions. J Biol Chem 275(8):5249–5252

Ishihara M, Kiefer MC, Barr PJ, Guo Y, Swiedler SJ (1992) Selection of cos cell mutants defective in the biosynthesis of heparan sulfate proteoglycan. Anal Biochem 206(2):400–407. doi:0003-2697 (92)90385-K [pii]

Ishihara M, Guo Y, Wei Z, Yang Z, Swiedler SJ, Orellana A, Hirschberg CB (1993) Regulation of biosynthesis of the basic fibroblast growth factor binding domains of heparan sulfate by heparan sulfate-n-deacetylase/n-sulfotransferase expression. J Biol Chem 268(27): 20091–20095

Jakobsson L, Kreuger J, Holmborn K, Lundin L, Eriksson I, Kjellen L, Claesson-Welsh L (2006) Heparan sulfate in trans potentiates vegfr-mediated angiogenesis. Dev Cell 10(5):625–634

Joyce JA, Freeman C, Meyer-Morse N, Parish CR, Hanahan D (2005) A functional heparan sulfate mimetic implicates both heparanase and heparan sulfate in tumor angiogenesis and invasion in a mouse model of multistage cancer. Oncogene 24(25):4037–4051. doi:1208602 [pii] 10.1038/sj.onc.1208602

Kharabi Masouleh B, Ten Dam GB, Wild MK, Seelige R, van der Vlag J, Rops AL, Echtermeyer FG, Vestweber D, van Kuppevelt TH, Kiesel L, Gotte M (2009) Role of the heparan sulfate proteoglycan syndecan-1 (cd138) in delayed-type hypersensitivity. J Immunol 182(8): 4985–4993. doi:182/8/4985 [pii] 10.4049/jimmunol.0800574

Kisanuki YY, Hammer RE, Miyazaki J, Williams SC, Richardson JA, Yanagisawa M (2001) Tie2-cre transgenic mice: a new model for endothelial cell-lineage analysis in vivo. Dev Biol 230 (2):230–242

Koenig A, Norgard-Sumnicht K, Linhardt R, Varki A (1998) Differential interactions of heparin and heparan sulfate glycosaminoglycans with the selectins. Implications for the use of unfractionated and low molecular weight heparins as therapeutic agents. J Clin Invest 101(4): 877–889

Kragh M, Loechel F (2005) Non-anti-coagulant heparins: a promising approach for prevention of tumor metastasis (review). Int J Oncol 27(4):1159–1167

Ledin J, Staatz W, Li JP, Gotte M, Selleck S, Kjellen L, Spillmann D (2004) Heparan sulfate structure in mice with genetically modified heparan sulfate production. J Biol Chem 279(41): 42732–42741

Lee DY, Park K, Kim SK, Park RW, Kwon IC, Kim SY, Byun Y (2008) Antimetastatic effect of an orally active heparin derivative on experimentally induced metastasis. Clin Cancer Res 14 (9):2841–2849. doi:14/9/2841 [pii] 10.1158/1078-0432.CCR-07-0641

Leppanen A, White SP, Helin J, McEver RP, Cummings RD (2000) Binding of glycosulfopeptides to p-selectin requires stereospecific contributions of individual tyrosine sulfate and sugar residues. J Biol Chem 275(50):39569–39578. doi:10.1074/jbc.M005005200 M005005200 [pii]

Li Q, Park PW, Wilson CL, Parks WC (2002) Matrilysin shedding of syndecan-1 regulates chemokine mobilization and transepithelial efflux of neutrophils in acute lung injury. Cell 111(5):635–646

Ludwig RJ, Alban S, Bistrian R, Boehncke WH, Kaufmann R, Henschler R, Gille J (2006a) The ability of different forms of heparins to suppress p-selectin function in vitro correlates to their inhibitory capacity on bloodborne metastasis in vivo. Thromb Haemost 95(3):535–540. doi:06030535 [pii] 10.1267/THRO06030535

Ludwig RJ, Alban S, Boehncke WH (2006b) Structural requirements of heparin and related molecules to exert a multitude of anti-inflammatory activities. Mini Rev Med Chem 6(9): 1009–1023

Luster AD, Greenberg SM, Leder P (1995) The ip-10 chemokine binds to a specific cell surface heparan sulfate site shared with platelet factor 4 and inhibits endothelial cell proliferation. J Exp Med 182(1):219–231

Massena S, Christoffersson G, Hjertstrom E, Zcharia E, Vlodavsky I, Ausmees N, Rolny C, Li JP, Phillipson M (2010) A chemotactic gradient sequestered on endothelial heparan sulfate induces directional intraluminal crawling of neutrophils. Blood 116:1924–1931. doi:blood-2010-01-266072 [pii] 10.1182/blood-2010-01-266072

McQuade KJ, Beauvais DM, Burbach BJ, Rapraeger AC (2006) Syndecan-1 regulates alphavbeta5 integrin activity in b82l fibroblasts. J Cell Sci 119(Pt 12):2445–2456

Middleton J, Neil S, Wintle J, Clark-Lewis I, Moore H, Lam C, Auer M, Hub E, Rot A (1997) Transcytosis and surface presentation of il-8 by venular endothelial cells. Cell 91(3):385–395

Mocco J, Shelton CE, Sergot P, Ducruet AF, Komotar RJ, Otten ML, Sosunov SA, Macarthur RB, Kennedy TP, Connolly ES Jr (2007) O-desulfated heparin improves outcome after rat cerebral ischemia/reperfusion injury. Neurosurgery 61(6):1297–1303; discussion 1303–1294. doi:10.1227/01.neu.0000306109.55174.e6 00006123-200712000-00021 [pii]

Nelson RM, Cecconi O, Roberts WG, Aruffo A, Linhardt RJ, Bevilacqua MP (1993) Heparin oligosaccharides bind l- and p-selectin and inhibit acute inflammation. Blood 82(11): 3253–3258

Norgard-Sumnicht K, Varki A (1995) Endothelial heparan sulfate proteoglycans that bind to l-selectin have glucosamine residues with unsubstituted amino groups. J Biol Chem 270(20): 12012–12024

Norgard-Sumnicht KE, Varki NM, Varki A (1993) Calcium-dependent heparin-like ligands for l-selectin in nonlymphoid endothelial cells. Science 261(5120):480–483

Parish CR (2006) The role of heparan sulphate in inflammation. Nat Rev Immunol 6(9):633–643. doi:nri1918 [pii] 10.1038/nri1918

Park PW, Pier GB, Preston MJ, Goldberger O, Fitzgerald ML, Bernfield M (2000a) Syndecan-1 shedding is enhanced by lasa, a secreted virulence factor of pseudomonas aeruginosa. J Biol Chem 275(5):3057–3064

Park PW, Reizes O, Bernfield M (2000b) Cell surface heparan sulfate proteoglycans: selective regulators of ligand-receptor encounters. J Biol Chem 275(39):29923–29926. doi:10.1074/jbc.R000008200 R000008200 [pii]

Remick DG, Green LB, Newcomb DE, Garg SJ, Bolgos GL, Call DR (2001) Cxc chemokine redundancy ensures local neutrophil recruitment during acute inflammation. Am J Pathol 159 (3):1149–1157

Ringvall M, Ledin J, Holmborn K, van Kuppevelt T, Ellin F, Eriksson I, Olofsson AM, Kjellen L, Forsberg E (2000) Defective heparan sulfate biosynthesis and neonatal lethality in mice lacking n-deacetylase/n-sulfotransferase-1. J Biol Chem 275(34):25926–25930

Risau W (1997) Mechanisms of angiogenesis. Nature 386(6626):671–674

Rollins BJ (1997) Chemokines. Blood 90(3):909–928

Rot A, von Andrian UH (2004) Chemokines in innate and adaptive host defense: Basic chemokinese grammar for immune cells. Annu Rev Immunol 22:891–928

Sasisekharan R, Shriver Z, Venkataraman G, Narayanasami U (2002) Roles of heparan-sulphate glycosaminoglycans in cancer. Nat Rev Cancer 2(7):521–528

Springer TA (1994) Traffic signals for lymphocyte recirculation and leukocyte emigration: the multistep paradigm. Cell 76(2):301–314

Stappenbeck TS, Hooper LV, Gordon JI (2002) Developmental regulation of intestinal angiogenesis by indigenous microbes via paneth cells. Proc Natl Acad Sci USA 99(24):15451–15455

Stevenson JL, Varki A, Borsig L (2007) Heparin attenuates metastasis mainly due to inhibition of p- and l-selectin, but non-anticoagulant heparins can have additional effects. Thromb Res 120 (Suppl 2):S107–S111. doi:S0049-3848(07)70138-X [pii] 10.1016/S0049-3848(07)70138-X

Tanaka Y, Kimata K, Wake A, Mine S, Morimoto I, Yamakawa N, Habuchi H, Ashikari S, Yamamoto H, Sakurai K, Yoshida K, Suzuki S, Eto S (1996) Heparan sulfate proteoglycan on leukemic cells is primarily involved in integrin triggering and its mediated adhesion to endothelial cells. J Exp Med 184(5):1987–1997

Tyrrell DJ, Horne AP, Holme KR, Preuss JM, Page CP (1999) Heparin in inflammation: potential therapeutic applications beyond anticoagulation. Adv Pharmacol 46:151–208

Varki A (1994) Selectin ligands. Proc Natl Acad Sci USA 91(16):7390–7397

Varki NM, Varki A (2002) Heparin inhibition of selectin-mediated interactions during the hematogenous phase of carcinoma metastasis: rationale for clinical studies in humans. Semin Thromb Hemost 28(1):53–66

Wang L, Brown JR, Varki A, Esko JD (2002) Heparin's anti-inflammatory effects require glucosamine 6-o-sulfation and are mediated by blockade of l- and p-selectins. J Clin Invest 110(1):127–136

Wang L, Fuster M, Sriramarao P, Esko JD (2005) Endothelial heparan sulfate deficiency impairs l-selectin- and chemokine-mediated neutrophil trafficking during inflammatory responses. Nat Immunol 6(9):902–910

Wei M, Tai G, Gao Y, Li N, Huang B, Zhou Y, Hao S, Zeng X (2004) Modified heparin inhibits p-selectin-mediated cell adhesion of human colon carcinoma cells to immobilized platelets under dynamic flow conditions. J Biol Chem 279(28):29202–29210. doi:10.1074/jbc.M312951200 M312951200 [pii]

Wei M, Gao Y, Tian M, Li N, Hao S, Zeng X (2005) Selectively desulfated heparin inhibits p-selectin-mediated adhesion of human melanoma cells. Cancer Lett 229(1):123–126. doi:S0304-3835(05)00111-4 [pii] 10.1016/j.canlet.2005.01.034

Witt DP, Lander AD (1994) Differential binding of chemokines to glycosaminoglycan subpopulations. Curr Biol 4(5):394–400

Young E (2008) The anti-inflammatory effects of heparin and related compounds. Thromb Res 122(6):743–752. doi:S0049-3848(07)00283-6 [pii] 10.1016/j.thromres.2006.10.026

Yuan L, Geng JG, Xu Q (2003) N-desulfated heparin improves concanavalin a-induced liver injury partly through inhibiting t lymphocyte adhesion. Inflamm Res 52(9):383–389. doi:10.1007/s00011-003-1190-8

Zhou T, Chen JL, Song W, Wang F, Zhang MJ, Ni PH, Geng JG (2002) Effect of n-desulfated heparin on hepatic/renal ischemia reperfusion injury in rats. World J Gastroenterol 8(5): 897–900

Zhou Z, Wang J, Cao R, Morita H, Soininen R, Chan KM, Liu B, Cao Y, Tryggvason K (2004) Impaired angiogenesis, delayed wound healing and retarded tumor growth in perlecan heparan sulfate-deficient mice. Cancer Res 64(14):4699–4702. doi:10.1158/0008-5472.CAN-04-0810 64/14/4699 [pii]

Chapter 2
Heparan Sulfate Proteoglycans in Infection

Allison H. Bartlett and Pyong Woo Park

Abstract To cause infections, microbial pathogens elaborate a multitude of factors that interact with host components. Using these host–pathogen interactions to their advantage, pathogens attach, invade, disseminate, and evade host defense mechanisms to promote their survival in the hostile host environment. Many viruses, bacteria, and parasites express adhesins that bind to cell surface heparan sulfate proteoglycans (HSPGs) to facilitate their initial attachment and subsequent cellular entry. Some pathogens also secrete virulence factors that modify HSPG expression. HSPGs are ubiquitously expressed on the cell surface of adherent cells and in the extracellular matrix. HSPGs are composed of one or several heparan sulfate (HS) glycosaminoglycan chains attached covalently to specific core proteins. For most intracellular pathogens, cell surface HSPGs serve as a scaffold that facilitates the interaction of microbes with secondary receptors that mediate host cell entry. Consistent with this mechanism, addition of HS or its pharmaceutical functional mimic, heparin, inhibits microbial attachment and entry into cultured host cells, and HS-binding pathogens can no longer attach or enter cultured host cells whose HS expression has been reduced by enzymatic treatment or chemical mutagenesis. In pathogens where the specific HS adhesin has been identified, mutant strains lacking HS adhesins are viable and show normal growth rates, suggesting that the capacity to interact with HSPGs is strictly a virulence activity. The goal of this chapter is to provide a mechanistic overview of our current understanding of how certain microbial pathogens subvert HSPGs to promote their infection, using specific HSPG–pathogen interactions as representative examples.

A.H. Bartlett
Department of Pediatrics, Baylor College of Medicine, Houston, TX 77030, USA

P.W. Park (✉)
Division of Respiratory Diseases, Children's Hospital, Harvard Medical School, 320 Longwood Avenue, Enders-461, Boston, MA 02115, USA
e-mail: pyong.park@childrens.harvard.edu

M.S.G. Pavão (ed.), *Glycans in Diseases and Therapeutics*, Biology of Extracellular Matrix, 31
DOI 10.1007/978-3-642-16833-8_2, © Springer-Verlag Berlin Heidelberg 2011

2.1 Primer on HSPG Biology

The complex interplay between hosts and pathogens has many common themes. One of the first steps in infection is attachment to host tissues. The expression pattern of host cell surface proteins that serve as receptors for pathogen attachment has a significant role in determining the initiation, progression, outcome, and tissue tropism of infections. Among these, HSPGs are expressed ubiquitously on the surface of adherent cells and in the extracellular matrix (ECM), and many viral, bacterial, and parasitic pathogens have been described to interact with HSPGs (Rostand and Esko 1997; Bernfield et al. 1999; Spillmann 2001; Chen et al. 2008). HSPGs bind to and regulate growth factors, cytokines, and chemokines, and ECM components (Bernfield et al. 1999; Park et al. 2000; Bartlett et al. 2008) and are known to be involved in processes as diverse as wound healing (Fears and Woods 2006; Alexopoulou et al. 2007), angiogenesis (Stringer 2006), and neuronal development (Reizes et al. 2008). HSPGs are composed of a protein core to which one or more heparan sulfate (HS) glycosaminoglycan (GAG) chains are covalently attached. HS chains can be both *N*- and *O*-sulfated, and it is with these chains that most pathogens interact. The highly complex mechanisms of HSPG biosynthesis have been partially defined and are discussed in detail in other chapters of this book.

The major ECM HSPGs, perlecan and agrin, help to form the structure of the basement membrane (BM) and modulate growth factors to affect cell survival, motility, and tissue morphogenesis (Iozzo 2005). The major cell surface HSPGs are syndecans and glypicans (Bernfield et al. 1999). There are four syndecans in mammals (syndecan-1 through -4) with distinct extracellular domains to which HS chains are attached distally to the plasma membrane, and highly conserved transmembrane and cytoplasmic domains. Syndecan cytoplasmic domains contain one invariant serine and three invariant tyrosine residues as well as a C-terminal Glu–Phe–Tyr–Ala PDZ binding domain. Syndecan cytoplasmic domains can regulate interactions with host proteins such as c-Src (Kinnunen et al. 1998), cortactin (Kinnunen et al. 1998), syntenin (Grootjans et al. 1997), protein kinase A (Hayashida et al. 2006), CASK/LIN2A (Cohen et al. 1998; Hsueh et al. 1998), and protein kinase Cα (Kessler et al. 1997), among other signaling and scaffolding proteins.

There are six glypicans in mammals that differ from syndecans in several ways. Syndecan core proteins are type 1 transmembrane proteins, whereas glypicans are covalently linked to the cell surface by glycosylphosphatidylinositol (GPI) anchors (David et al. 1990). The extracellular core proteins of syndecans are rich in secondary structure-breaking proline residues and considered linear, whereas glypicans are thought to have a compact globular structure held in place by multiple disulfide bonds formed between several conserved cysteine residues (Chen and Lander 2001). Further, HS attachment sites in syndecans are distal to the plasma membrane, whereas those of glypicans are proximal (Chen and Lander 2001). In addition, syndecans and glypicans show distinct temporal and spatial expression patterns, which in part explain how these HSPGs may function specifically in vivo.

Cell surface HSPGs primarily function as coreceptors for HS-binding ligands, serving as a scaffold that localizes the ligands to the cell surface, enabling them to interact more efficiently with their respective signaling receptors. Cell surface HSPGs can also function as a primary receptor where they mediate the endocytosis of several HS-binding ligands via macropinocytosis and clathrin-dependent and clathrin-independent endocytosis (Poon and Gariepy 2007). Some of the examples include macropinocytotic uptake of FGF-2 (Tkachenko et al. 2004) and clathrin-independent endocytosis of unmodified lipoproteins through syndecans (Fuki et al. 1997). Cell surface HSPG-mediated endocytosis is thus an important mechanism that regulates growth factor signaling and lipoprotein degradation, though how cell surface HSPGs mediate outside-in signaling leading to ligand internalization is incompletely understood. As described later, certain microbial pathogens have adapted or evolved to subvert this mechanism to gain entry into host cells.

The syndecan family of cell surface HSPGs can be shed as intact, soluble HSPG ectodomains by proteolytic cleavage of the core protein. Soluble syndecan-1 and -4 ectodomains are detected in inflamed or infected body fluids, indicating a physiological role for shedding in diseases (Subramanian et al. 1997; Wang et al. 2008; Zvibel et al. 2009). Syndecan ectodomains are replete with all of their HS chains and are thought to maintain their ability to interact with the same ligands as the cell surface syndecans, and thus act as soluble autocrine or paracrine effectors. For example, syndecan-1 ectodomains regulate the proliferative response of cells in injured tissue to FGF-2 (Kato et al. 1998) and syndecan-1 ectodomains increase the invasiveness of a human breast cancer cell line in vitro (Nikolova et al. 2009). Syndecan-1 ectodomains also bind to neutrophil enzymes such as elastase and cathepsin G, protecting the enzymes from their physiological inhibitors and potentiating the enzymes' activity (Kainulainen et al. 1998).

Our understanding of the mechanism whereby syndecan shedding occurs and is regulated is increasing. Syndecan ectodomains are shed by metalloproteinase sheddases. For example, matrix metalloproteinase-7 (MMP-7, matrilysin) (Li et al. 2002; Ding et al. 2005), MMP-9 (gelatinase B) (Brule et al. 2006), MMP-14 (MT1-MMP) (Endo et al. 2003), and ADAM17 (TACE) (Pruessmeyer et al. 2010) can shed syndecan-1 ectodomains. Chemical inhibitor studies have shown that several kinases regulate syndecan shedding, such as MAP kinases ERK (extracellular signal-related kinase) and JNK (c-Jun NH_2-terminal kinase) (Fitzgerald et al. 2000), protein kinase C (PKC) (Subramanian et al. 1997; Fitzgerald et al. 2000), and protein tyrosine kinases (PTKs) (Fitzgerald et al. 2000; Park et al. 2000, 2004; Chung et al. 2006). The role of the highly conserved cytoplasmic domain of syndecan-1 in shedding regulation is becoming clear. The three invariant Tyr residues are required for agonist-induced syndecan-1 shedding, but Tyr phosphorylation of the cytoplasmic domain is not (Hayashida et al. 2008). Instead the cytoplasmic domain modulates syndecan-1 shedding by interacting with critical regulators of shedding. One such regulator identified recently is Rab5. Rab5 is a small GTP-binding protein, whose GDP-bound inactive form binds preferentially to the cytoplasmic domain of syndecan-1 (Hayashida et al. 2008). Stimulation of cells with syndecan-1 shedding agonists induces the activation of Rab5 and dissociation of activated GTP-bound Rab5 from the syndecan-1 cytoplasmic domain, suggesting that Rab5 regulates syndecan-1 shedding by serving as an on–off switch of shedding through its alternation between GDP- and GTP-bound states (Hayashida et al. 2008).

Shed glypican ectodomains have been detected in cell culture media (Mertens et al. 1996; Fitzgerald et al. 2000), suggesting that glypicans are also shed. Indeed, the Drosophila Dally (glypican)-like protein Dlp can be cleaved from its GPI anchor and released by the protein Notum (a secreted member of the α/β-hydrolase family) altering the activity of the growth factor wingless (Wg) (Kreugera et al. 2004). Further, glypican-3 ectodomains have been detected in the serum of patients with hepatocellular carcinoma (Capurro et al. 2003; Hippo et al. 2004) and melanoma (Nakatsura et al. 2004), but whether this represents shedding or secretion remains uncertain. One study identified the soluble glypican-3 fragment in patients with hepatocellular carcinoma as the N-terminal end, with the core protein cleaved between ^{358}Arg and ^{359}Ser (Hippo et al. 2004), suggesting physiologic shedding of glypican-3 in the context of hepatocellular carcinoma. However, whether glypicans are shed upon infectious challenge remains to be determined.

2.2 Methods for HSPG Studies in Infectious Diseases

Several techniques have been applied to investigate the role of HSPGs in infections. One of the simplest methods to determine if HSPGs are involved in a process is to evaluate the ability of soluble HS or heparin to specifically inhibit the function being studied (e.g., attachment, internalization) (Rostand and Esko 1997). Comparing the ability of different GAGs to inhibit attachment also gives information on which proteoglycan(s) might be involved. When a specific HSPG, such as syndecan-1 is suspected, soluble forms of the purified ectodomain can also be used in addition to HS and heparin to probe the specific interaction. Further, various chemically desulfated heparin compounds have been used to determine the critical HS modification in HS–pathogen interactions. In addition, recent studies have generated means to produce structurally defined HS using specific HS biosynthetic enzymes in vitro (Chen et al. 2007a), which should further advance our understanding of critical HS modifications in future studies.

Additional information regarding precise GAG involvement can be obtained by treating cells in culture with a variety of GAG-digesting enzymes. There are three types of *Flavobacterium* heparinases (I, II, and III) with different substrate specificities and three types of chondroitinases (ABC, AC, and C) that allow one to determine the relative contribution of different GAGs by selective removal of GAGs. For example, attachment of *Toxoplasma gondii* to Vero cells is inhibited by heparinase I and II treatment, but unaffected by chondroitinase ABC treatment, suggesting that sulfated domains in HS mediate *T. gondii* attachment. In contrast, the attachment of *Neospora caninum* tachyzoites, a close relative of *T. gondii*, to Vero cells is unaffected by heparinase treatment and inhibited by chondroitinase treatment

(Naguleswaran et al. 2002), suggesting that *N. caninum* tachyzoites bind to CS to adhere to Vero cells.

Selective chemical inhibitors of HS biosynthesis, such as sodium chlorate and β-D-xylosides, have also been used to study the specificity and significance of HS–pathogen interactions. Growing cells in media that is deficient in inorganic sulfate and supplemented with the sulfation inhibitor sodium chlorate reduces the overall extent of sulfation of HS and other GAGs (Baeuerle and Huttner 1986). β-D-xylosides competitively inhibit the formation of the tetrasaccharide linkage region required for GAG attachment to proteoglycan core proteins, diverting the machinery and raw materials from building GAG chains on core proteins. The GAG chains built on the β-D-xylosides primers are secreted from cells, and the proteoglycans displayed on the cell surface are underglycanated (Esko and Montgomery 1995). However, caution must be taken when using sodium chlorate and β-D-xylosides because both can be toxic to cells when incubated at high concentrations or for long periods.

Cell lines deficient in certain HS biosynthetic enzymes have also been used as tools to study the relationship between the type and degree of sulfation in the context of microbial infection. Multiple mutant CHO K1 cell lines exist which produce different amounts and types of HS and CS: pgsA-745, which lacks xylosyltransferase produces no GAGs; pgsB-761, -618, and -650 have defects in galactosyltransferase I and produce ~5%, ~15%, or ~30% of wild type HS and CS; pgsD-677 has defects in *N*-acetylglucosaminyl transferase and glucuronosyltransferase and makes three times the wild type amount of CS but cannot make HS; pgsF-17 lacks functional 2-*O*-sulfotransferase thus produces 2-*O*-sulfate deficient HS but normal CS (Rostand and Esko 1997). Several specific sulfatases have also been identified, and these should also facilitate the determination of critical sulfate modifications in microbial pathogenesis. A method for synthesizing HS oligosaccharides with defined composition has been developed for studying structure–activity relations of specific HS oligosaccharides with HS-binding proteins (Arungundram et al. 2009). This method has not yet, to our knowledge, been used in the context of evaluating host–pathogen interactions, but remains an exciting tool likely to be used in the future to enhance our understanding of the molecular interactions between pathogens and HS. Lastly, recent studies have generated mutant mouse lines deficient in certain HS biosynthetic enzymes (Forsberg et al. 1999; Grobe et al. 2005; Pallerla et al. 2008) and these mice should allow investigators to better examine the physiological significance of HS modifications in infectious diseases.

2.3 HSPGs in Microbial Attachment and Internalization

Many viral, bacterial, and parasitic pathogens have been shown to subvert HSPGs in the course of infection (Table 2.1). In microbial pathogenesis, HSPGs function primarily as initial, low affinity coreceptors that concentrate pathogens on host cell surfaces, increasing binding to specific secondary receptors. For several pathogens

Table 2.1 HSPG–pathogen interactions

Pathogen	Pathogen protein	HSPG	Function/ interaction	Reference
Bacteria				
Bacillus anthracis	AnlB, ANIO, InhA, Npr599	Syndecan-1	Shedding	Chung et al. (2006), Popova et al. (2006)
Bacillus cereus	ClnA	Syndecan-1	Shedding	Popova et al. (2006)
Borrelia burgdorferi	39 kDa protein	Unknown	Attachment	Isaacs (1994)
Bordetella pertussis	Filamentous hemagglutinin	Unknown	Attachment	Hannah et al. (1994)
Chlamydia pneumoniae	OmcB	Unknown	Attachment	Moelleken and Hegemann (2008)
Chlamydia trachomatis	Unknown	Unknown	Attachment, invasion	Zhang and Stephens (1992), Chen and Stephens (1997)
Haemophilus influenzae, nontypable	High molecular weight protein (HMW)	Unknown	Attachment	Noel et al. (1994)
Helicobacter pylori	Vacuolating cytotoxin (VacA)	Unknown	Toxin internalization	Utt et al. (2001)
Listeria monocytogenes	ActA	Syndecan-1	Attachment, invasion	Alvarez-Dominguez et al. (1997)
Mycobacterium tuberculosis	Hemagglutinin	Unknown	Attachment	Pethe et al. (2001)
Neisseria gonorrhoaea	Opa	Syndecan-1, -4	Attachment, invasion	Freissler et al. (2000)
Neisseria meningitidis	Opc	Unknown	Attachment, invasion	de Vries et al. (2002)
Neisseria meningitidis	GNA2132 (Neisserial Heparin Binding Antigen (NHBA))	Unknown	(presumptive) Attachment, resistance to serum killing	Serruto et al. (2010)
Orientia tsutsugamushi	Unknown	Syndecan-4	Attachment, invasion	Kim et al. (2004)
Porphyromonas gingivalis	LPS, gingipains	Syndecan-1	Shedding	Andrian et al. (2005, 2006)
Pseudomonas aeruginosa	LasA	Syndecan-1	Shedding	Park et al. (2000)
Staphylococcus aureus	α-toxin, β-toxin	Syndecan-1	Shedding	Park et al. (2004)
Streptococcus agalactiae	Alpha C protein	Unknown	Attachment, invasion	Baron et al. (2004)
Streptococcus pyogenes	M protein	Unknown	Attachment	Frick et al. (2003)
Streptococcus pneumoniae	ZmpC	Syndecan-1	Shedding	Chen et al. (2007b)
Yersinia enterocolitica	LcrG	Unknown	Attachment, invasion	Boyd et al. (1998)
Viruses				
Adeno-associated virus type 2	Capsid protein VP3	Unknown	Attachment	Opie et al. (2003)
Adenovirus	Ad3 Fiber knob	Unknown	Attachment	Tuve et al. (2008)
Coronavirus	Spike protein	Unknown	Attachment	de Haan et al. (2008)
Coxsackievirus	Capsid protein VP1	N- and 6-0-sulfated HSPGs	Attachment, endocytosis	Zautner et al. (2003, 2006)

(*continued*)

Table 2.1 (continued)

Pathogen	Pathogen protein	HSPG	Function/ interaction	Reference
Cytomegalovirus	gB	Unknown	Attachment	Boyle and Compton (1998)
Dengue virus	E (envelope protein)	Unknown	Attachment, internalization	Chen et al. (1997), Hilgard and Stockert (2000)
FMDV	VP3	Unknown	Attachment	Fry et al. (1999)
HSV-1 and -2	gB, gC, gD	Syndecan-2	Attachment	Spear (2004), Cheshenko et al. (2007)
Hepatitis B virus	Large viral envelope protein	Unknown	Attachment	Schulze et al. (2007)
Hepatitis C virus	Envelope glycoprotein E2	Unknown	Attachment	Barth et al. (2006)
HHV-8 (KSHV)	gB, gpK8.1A	Unknown	Attachment	Akula et al. (2001), Veeltil et al. (2006)
HIV-1	Tat	Perlecan	Tat internalization	Argyris et al. (2004)
	Tat	Unknown	Lymphoid cell extravasation	Urbinati et al. (2009)
	gp120	Syndecan-3	Attachment	de Witte et al. (2007)
	gp41	Agrin	Attachment	Alfsen et al. (2005)
HPV	L1 carboxy terminal	Syndecan-1, -3, -4, glypican-1	Attachment	Joyce et al. (1999), Shafti-Keramat et al. (2003), de Witte et al. (2008)
HTLV1	Surface glycoprotein gp46	Unknown	Attachment	Piñon et al. (2003)
Japanese encephalitis virus	Envelope (E) protein	Unknown	Attachment	Lee et al. (2004)
Pseudorabies virus	Glycoprotein C	Unknown	Attachment	Trybala et al. (1998)
Respiratory syncytial virus	Fusion glycoprotein	Unknown	Attachment, infectivity	Crim et al. (2007)
Rhinovirus	VP1	Unknown	Attachment	Vlasak et al. (2005)
Sindbis virus	E2 envelope glycoprotein	Unknown	Attachment	Ryman et al. (2007)
Vaccinia virus	Viral envelope protein A27L	Unknown	Fusion	Hsiao et al. (1998)
	VCP (Vaccinia virus complement control protein)	Unknown	Anchoring of RCA (regulators of complement activation) to cell membrane	Murthy et al. (2001)
West Nile virus	Envelope (E) protein	Unknown	Attachment	Lee et al. (2004)
Yellow fever virus	Envelope (E) protein	Unknown	Attachment	Nickells et al. (2008)
Parasites				
Giardia lamblia	Alpha-1 giardin	Unknown	Attachment	Weiland et al. (2003)
Leishmania spp.	Unknown	Unknown	Attachment	Love et al. (1993)
Encephalitozoon spp.	Spore wall protein EnP1	Unknown	Attachment	Southern et al. (2007)

(*continued*)

Table 2.1 (continued)

Pathogen	Pathogen protein	HSPG	Function/ interaction	Reference
Neospora caninum	Microneme protein $NcMIC_3$	Unknown (CS)	Attachment	Naguleswaran et al. (2002)
Plasmodium spp.	Circumsporozoite protein (CSP)	Multiple	CSP cleavage, productive invasion	Coppi et al. (2007)
	Thrombospondin-related anonymous protein (TRAP)	Unknown	Invasion	McCormick et al. (1999)
	BAEBL (EBA-140)	Multiple (HS)	Invasion	Kobayashi et al. (2010)
	VAR2CSA	Chondroitin Sulfate A (CSA)	Attachment (to placenta)	Khunrae et al. (2010)
Toxoplasma gondii	Surface antigen 3	Unknown	Attachment	Jacquet et al. (2001)
	Unknown	Unknown (*N*-sulfation required)	Replication in parasitophorous vesicle	Bishop et al. (2005)
	Programmed cell death 5 (TgPDCD5)		Protein internalization and enhanced apoptotic activity	Bannai et al. (2008)
Trypanosoma cruzi	Cruzipain	HSPG	Enhanced enzymatic activity	Lima et al. (2002)
	Heparin-binding protein (HBP-Tc)	Unknown	Attachment	Oliveira et al. (2008)
Other				
Prion	PrP^{C}	Glypican-1	Lipid raft association, conversion to PrP^{Sc}	Horonchik et al. (2005), Taylor et al. (2009)

and virulence factors, HSPGs can also function as a direct internalization receptor. Further, soluble HSPG ectodomains can enhance bacterial virulence in vivo by inhibiting innate host defense mechanisms. The following sections describe these major mechanisms of HSPG subversion using specific pathogens as examples. Other pathogens that use similar mechanisms for their pathogenesis are summarized in Table 2.1, and include *Bordetella pertussis*, *Mycobacterium tuberculosis*, cytomegalovirus, and *Giardia lamblia*, among many others. A prototype of a pathogen using HSPGs as an initial attachment receptor is shown in Fig. 2.1a.

2.3.1 Papillomaviruses

Papillomaviruses are nonenveloped, double-stranded DNA viruses that infect only skin and mucosal epithelial cells and manipulate the host cell cycle to create an environment ideal for viral replication. This alteration of the host cell cycle also

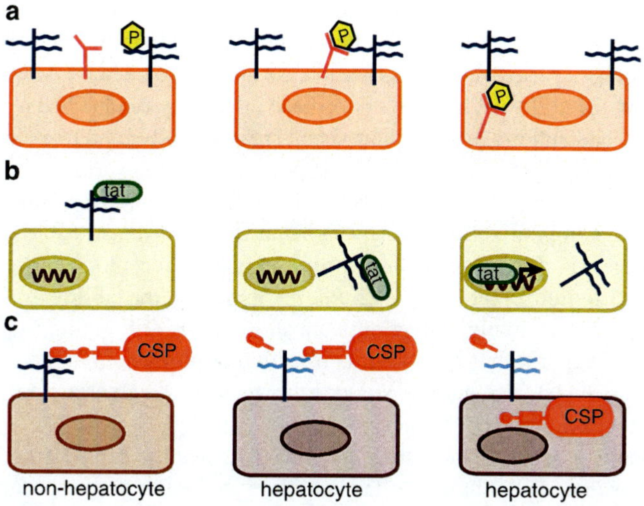

Fig. 2.1 Mechanisms of microbial subversion of cell surface HSPGs. (**a**) A pathogen (P) binds to a cell surface HSPG which then facilitates binding of the pathogen to its specific receptor (*shown in pink*) leading to internalization of the pathogen–receptor complex. (**b**) HIV Tat protein binds to cell surface HSPGs and is internalized. Tat then travels to the nucleus where it transactivates transcription of host cell genes. (**c**) *P. falciparum* circumsporozoite protein (CSP) binds to HSPGs (*dark blue*). When it encounters a highly sulfated HSPG (*light blue*) as is found in hepatocytes, CSP is cleaved and the remaining molecule is internalized

results in formation of squamous fibroepitheilial tumors, condyloma, and malignant epithelial tumors. Over 100 types of human papillomaviruses (HPV) have been identified. HPVs are separated into two genuses: alpha, containing strains that preferentially infect the genital mucosa (including oncogenic strains 16 and 18 which cause cervical cancer) and beta, containing strains that cause skin infections. Papillomavirus capsids are made of a major capsid protein, L1, and a minor capsid protein, L2. Although papillomaviruses cannot be propagated in vitro, several substitute tools exist. Virus-like particles (VLPs) are generated by the synthesis of capsid proteins L1 and L2 (Kirnbauer et al. 1992; Rose et al. 1993; Volpers et al. 1994) and can bind to most cell lines tested (Qi et al. 1995; Volpers et al. 1995). Pseudovirions, VLPs with a marker plasmid inside, were created to facilitate the study of attachment and invasion (Unckell et al. 1997).

Research on HPV types 11 (Joyce et al. 1999), 16, 33, and 39 (Giroglou et al. 2001) showed a key role for HS in pseudoinfection. Heparin, but not dermatan sulfate or CS, blocked the attachment of pseudovirions to COS-7 cells (Giroglou et al. 2001). In agreement with this, treatment of cells with heparinase also blocked the ability of pseudovirions to attach (Giroglou et al. 2001). As is the case for other pathogens, HSPGs serve as low-affinity, but abundant, primary attachment receptors and a secondary receptor mediates internalization of HPVs. Christensen et al. (1995) found that neutralizing antibodies against VLPs were effective in preventing internalization up to 8 h after infection – without affecting the ability of the VLPs to attach to host cells (Christensen et al. 1995). Giroglou et al. (2001) extended the understanding of the timing of attachment and internalization: both neutralizing antibodies and heparin were able to decrease infectivity when added early after infection, but the ability of heparin to decrease infectivity diminished over a shorter time course. The authors propose that virions may initially bind to a single proteoglycan – from which they can easily be displaced by free heparin. Over time, more proteoglycan molecules are recruited and bind to each virion, making it less susceptible to competitive detachment by heparin, but still accessible to neutralizing antibodies which block attachment to a secondary receptor required for internalization (Giroglou et al. 2001). Studies comparing the ability of chemically modified heparins to inhibit infection have shown that VLP binding requires only *O*-sulfation of heparin, but pseudovirus infection requires both *N*- and *O*-sulfation (Selinka et al. 2003). The authors suggest this may be due to structural changes in the capsid proteins after DNA encapsidation (Selinka et al. 2003).

Continued studies are elucidating the specific roles for capsid proteins L1 and L2. HSPG-mediated attachment of L1 to a variety of cell types is an important first step in establishing infection. Following this initial binding, it is thought that the minor capsid protein L2 binds to a different receptor and mediates internalization. Consistent with this mechanism, it has been shown that virions with L1 and L2 have increased infectivity over virions with only L1 (Unckell et al. 1997; Yeager et al. 2000; Roden et al. 2001). In addition, neither anti-L2 neutralizing antibody (Roden et al. 1994a, b, 1995) nor deletion of either the N- or C-terminus of L2 can block virion binding to the host cell surface (Yang et al. 2003). In the case of HPV16, the L2 residues responsible for facilitating infection (aa residues 13–31) are not displayed on the capsid surface (Kawana et al. 1998, 1999), whereas a neighboring series of residues (aa residues 32–81) are present (Heino et al. 1995; Liu et al. 1997). This suggests a model in which HSPG binding of capsid protein L1 leads to a conformational change exposing residues on L2 that can then bind a secondary receptor to enhance infectivity. Richards et al. (2006) identified a host enzyme responsible for cleavage of the N-terminal residues, the proprotein convertase furin, which acts after the conformational change has occurred (Richards et al. 2006). The importance of this conformational change was highlighted by reports that L1-specific neutralizing antibodies induced by vaccination with HPV 16-VLP do not prevent attachment of mature virus to the host cell surface, but rather inhibit infection by preventing the conformational change of L2 (Day et al. 2008).

The secondary receptors that mediate papillomavirus internalization are not known. Using a related virus, bovine papillomavirus (BPV), Bossis et al. (2005) identified a series of conserved residues in the minor capsid protein L2 that are critical for association with the snare syntaxin 18 and subsequent trafficking to the endoplasmic reticulum (ER). The relevance of this interaction between capsid protein L2 and syntaxin 18 in endocytosis of HPV is unclear. Patterson et al. (2005) showed that HPV 31b infection of untransformed human keratinocytes in vitro did not require HS, but HS was required for HPV 31b infection of other cell lines, such as COS-7. The authors speculate that perhaps the secondary high-affinity receptor is so prevalent, that the primary, low-affinity HSPG interaction is not required for infection in the natural host cell, the keratinocyte, and cast doubt upon the relevance of other in vitro findings as they relate to what truly happens in vivo (Patterson et al. 2005).

In vitro studies of papillomaviruses are important, but the ability of these findings to predict in vivo mechanisms are limited and the true role of HSPGs in HPV infections in vivo is not certain. Johnson et al. (2009) developed a murine model of cervicovaginal challenge and investigated the role for HSPGs in HPV 5, 16, and 31 attachment in vivo (Johnson et al. 2009). Heparin was able to inhibit adsorption of HPV 16 and 31 but not HPV 5 in vivo (Buck et al. 2006; Johnson et al. 2009). Together, these studies suggest that certain strains of HPV require HSPGs for fulminant infection in vivo, but this remains to be rigorously tested. Recent in vivo studies suggest that the L2 capsid protein cleavage event occurs while the virion is bound to HSPG in the BM, and the subsequent conformational change decreases the VLP's affinity for HSPG, facilitating transfer to a non-HSPG receptor on the epithelial cell (Kines et al. 2009). The ECM produced by epithelial cells in vitro does not induce the same changes in the virion as BM in vivo, and can lead to underestimation of the importance of HSPG in HPV infection (Kines et al. 2009).

2.3.2 Dengue Virus

Dengue virus is a positive-sense, single-stranded RNA virus in the *Flavivirus* genus that is spread by mosquitoes. Each of the four serotypes can infect humans and cause diseases including dengue fever, dengue hemorrhagic fever, and dengue shock syndrome. The envelope (E) protein of dengue virus, one of three structural proteins of the virion, is able to bind HS both as a purified protein (Chen et al. 1997) and as a virion component (Germi et al. 2002). Heparin is also able to inhibit viral penetration as well as attachment (Hung et al. 1999). Several other proteins have also been identified as part of a receptor complex for dengue virus. GRP78 (BiP) was identified as a key receptor for internalization of dengue in hepatocytes (Jindadamrongwech et al. 2004). Heat shock proteins (HSP) 70 and 90 have also been identified as coreceptors for dengue. Upon incubation with E protein, HSP70 and 90 on host cell membranes relocate into lipid rafts (Reyes-del Valle et al. 2005). The integrity of these lipid rafts is critical for infection, suggesting that these rafts facilitate clustering of the dengue virus receptor complex (Reyes-del Valle et al. 2005). The relationship between HSPGs and the other identified receptors for dengue is controversial. That λ-carrageenan, a sulfated galactan that presumably mimics HS can block dengue attachment and internalization but not replication events after internalization (Talarico and Damonte 2007), and heat shock of cells increases infectivity of dengue virus, but does not affect attachment (Chavez-Salinas et al. 2008) supports a role for HS in initial attachment of virus to host cells. Experiments with carrageenans also showed a potential role for HS in viral uncoating into the host cell as well (Talarico and Damonte 2007), illustrating the multiple ways by which dengue virus subverts HSPGs to cause infection.

2.3.3 *Listeria monocytogenes*

L. monocytogenes is a Gram-positive, intracellular bacterial pathogen. A food-borne pathogen, it crosses the intestinal mucosa and enters the systemic circulation where it can cause meningitis and sepsis in immunocompromised hosts, including pregnant women and fetuses. *L. monocytogenes* gains entry to cells via multiple mechanisms. Phagocytic cell uptake of *L. monocytogenes* occurs via the C3bi (Drevets and Campbell 1991) and C1q (Alvarez-Dominguez et al. 1993) complement receptors. Epithelial cell uptake occurs when internalin protein A (InlA) binds to E-cadherin (Mengaud et al. 1996) and the bacterium is endocytosed via caveolin or clathrin (Bonazzi et al. 2008). Hepatocyte uptake occurs when Internalin B (InlB) binds to the hepatocyte growth factor receptor Met, causing mono-ubiquitination of Met and subsequent clathrin-dependent endocytosis (Veiga and Cossart 2005). InlB also binds HSPGs and potentiates the activity of InlB-Met binding, possibly by clustering InlB and concentrating Met receptors (Jonquières et al. 2001) or by stabilizing the InlB–Met complex during invasion (Banerjee et al. 2004). Lastly, nonphagocytic cell uptake can also occur when InlB binds the C1q complement receptor (gC1qR) leading to phosphorylation of the adaptor proteins Gab1, Cbl, and Shc (Braun et al. 2000).

While InlA and InlB are clearly important in *L. monocytogenes* pathogenesis, InlA and B deletion mutant strains are still capable of invasion, suggesting other mechanisms are available (Dramsi et al. 1995, 1997; Lingnau et al. 1995). Indeed, the *L. monocytogenes* surface protein ActA has been shown to have a role in adherence and invasion (Alvarez-Dominguez et al. 1997), although its best-described function is to manipulate the host cell cytoskeleton to allow bacterial migration within and between host cells (Portnoy et al. 1992; Sheehan et al. 1994; Lasa and Cossart 1996). Alvarez-Dominguez et al. (1997) demonstrated that *L. monocytogenes* ActA, with three regions of positively charged residues, mediates attachment to cell surface HSPGs. ActA appears to mediate attachment to and invasion of epithelial cells, possibly through microvilli (Suarez et al. 2001). Consistent with this, coinstillation of heparin during oral inoculation of mice with *L. monocytogenes* did not alter the bacterial load in the cecum, but did decrease extraintestinal dissemination, supporting the role of HSPGs in attachment and internalization of *L. monocytogenes* (Henry-Stanley et al. 2003). Although incompletely understood, the ability of ActA to interact with HSPGs and specifically mediate attachment of epithelial cells with microvilli may have important ramifications regarding how the bacterium crosses the intestinal epithelium.

2.4 HSPG as a Receptor for Virulence Factors

The HIV transcriptional activator protein (Tat) is, as its name suggests, the main transactivator of HIV (Gatignol and Jeang 2000). Tat is a small, cationic polypeptide that is released from HIV-infected cells (Noonan and Albini 2000) and can be detected in the serum of HIV-infected individuals (Westendorp et al. 1995). Tat has important roles in biological effects of HIV on non-CD4 cells. Tat is thought to be a neurotoxin important for the pathogenesis of AIDS dementia (Dewhurst et al. 1996). Tat is also involved in tumorigenesis by inducing Kaposi Sarcoma Herpes Virus replication (Zeng et al. 2007). Four classes of receptors are known to interact with Tat: αvβ3 integrin (Urbinati et al. 2005); VEGF receptors Flt-1 and Flk-1/KDR (Albini et al. 1996); chemokine receptors CCR2, CCR3, and CXCR4 (Albini et al. 1998; Xiao et al. 2000); and HSPGs (Rusnati and Presta 2002).

Tat binds to HSPGs through a region of basic residues, and can be released intact and active from HSPGs by treatment with heparinase (Chang et al. 1997). Tat binding to cell surface HSPGs facilitates its internalization (Tyagi et al. 2001) and subsequent activation of transcription (Fig. 2.1b). Using a GFP-Tat fusion protein, Tyagi et al. (2001) demonstrated that Tat uptake requires cell surface HSPGs. The interaction between HSPGs and Tat depends on the size of HS and degree and type of sulfation (Rusnati et al. 1997, 1999). Overproduction of CS cannot compensate for a lack of HS, demonstrating the functional specificity of the Tat–HSPG interaction (Tyagi et al. 2001). As is seen in HSPG–growth factor interactions, HSPG binding leads to Tat oligomerization (Rusnati et al. 1999) which facilitates tyrosine-kinase receptor dimerization and signaling. Heparin–Tat binding also protects Tat from proteolytic degradation (Chang et al. 1997). Although this does not exclude the involvement of other types of HSPGs, studies have shown that the cell line WiDr, which lacks all HSPGs except perlecan, is permissive for Tat internalization (Argyris et al. 2004).

The HSPG–Tat interaction has also been shown to be important for lymphoid cell extravasation during HIV infection (Urbinati et al. 2009), which leads to viral dissemination and AIDS-associated leukemia/lymphoma (Chirivi et al. 1999). Tat accumulates on endothelial cell (EC) and B lymphoid cell surfaces by binding HSPGs (Urbinati et al. 2009). Tat bound to HSPGs on ECs or B lymphoid cells significantly increases adhesion to B lymphoid cells or ECs, respectively. Interaction between EC and B lymphoid cells mediated by HSPG requires Tat homodimerization and leads to B lymphoid cell transendothelial migration and extravasation.

The HIV protein gp120 is the envelope glycoprotein, which binds to the host cell receptor CD4. gp120 can also be found in a "free" form and, like Tat, can bind to HSPGs. This can lead to ECM accumulation of the free form of gp120 bound to HSPGs (Klasse and Moore 2004). HSPG binding of the envelope-embedded gp120 is thought to increase concentration of viruses on the cell surface facilitating interactions with HIV receptors such as CD4, CCR5, and CXCR4 (Clapham and McKnight 2001) thereby increasing infectivity. The use of polyanionic compounds to inhibit HIV binding mediated by both Tat and gp120 has been explored, but successful use of these compounds as a therapy remains to be established (Bugatti et al. 2007). Three of the four heparin binding domains identified in gp120 are in domains involved in coreceptor recognition and are located near each other and near the coreceptor binding site, making this an important potential target for drug development (Crublet et al. 2008). A compound has recently been developed that takes advantage of the ability of gp120 to bind HS and a coreceptor. A CD4-mimetic peptide linked to a HS dodecasaccharide has been developed that binds to gp120, blocks the CD4 binding site, and opens the cryptic coreceptor binding domain to HS-mediated blocking (Baleux et al. 2009). In vitro, this compound effectively inhibits CCR5-, CXCR4-, and dual-tropic HIV-1, which is significant because currently there are no inhibitors available to block CXCR4 binding (Baleux et al. 2009).

The HIV–HS interaction is also implicated in the sexual transmission of HIV infection. HIV virions can bind to HS expressed by spermatozoa and thereby be transmitted in semen along with free virions and infected leukocytes (Ceballos et al. 2009). These spermatozoa-associated virions are then efficiently transmitted to dendritic cells (DCs), macrophages, and T-cells. Low pH (~6.5), similar to that of the vaginal mucosa after sexual intercourse, enhances transmission of virions to DCs, highlighting the potentially critical role in spermatozoa and DCs in HIV transmission in vivo (Ceballos et al. 2009). The source of HS on spermatozoa is most likely Sdc-3 and -4, as Sdc-1 and -2 are not expressed, and glypicans are not involved.

2.5 HSPGs in Tissue Tropism of Pathogens

Plasmodium falciparum sporozoites are the infectious agents that cause malaria, one of the most common infectious diseases worldwide. Infection occurs when an infected mosquito injects sporozoites into the skin of a mammalian host. The cycle that follows is complex, taking the sporozoite through the bloodstream to the liver where it invades hepatocytes and is transformed into extraerythrocytic forms. Eventually, each extraerythrocytic form releases thousands of merozoites. Merozoites enter the blood stream and infect erythrocytes, causing the symptoms of malarial disease, such as anemia, fever, arthralgia and in severe cases, coma and death.

The role of HSPGs in the *Plasmodium* life cycle is becoming clear and is unique among the pathogens discussed thus far (Fig. 2.1c). *Plasmodium* apparently uses the differential degree of sulfation of HSPGs in different tissues to complete its journey to the liver. The major surface protein of circumsporozoites, circumsporozoite protein (CSP), has been a target of many malaria vaccine efforts (Sharma and Pathak 2008). It attaches to hepatocytes through interaction with HSPGs, and this binding can be inhibited by heparinase treatment (Frevert et al. 1993). HSPGs are present on the surface of multiple cell types, but hepatocyte HSPGs are more sulfated than endothelial and dermal cell HSPGs (Lindblom and Fransson 1990; Lyon et al. 1994). Thus, it was hypothesized that the degree of sulfation, rather than simply the presence of HSPGs, was the key factor that allows sporozoites to migrate through multiple cell types and invade hepatocytes. Consistent with this, inhibition of sulfation decreased sporozoite migration (Humphries and Silbert 1988), and experiments of sporozoite migration in dermal fibroblasts, ECs, and hepatocytes showed that overall sulfation level of the cell type correlated inversely with migratory activity (Coppi et al. 2007).

Further study has shown that the HSPG–CSP interaction may regulate the proteolytic cleavage of CSP. CSP cleavage is associated with productive invasion (Coppi et al. 2005) and rapid cleavage is induced by highly sulfated HS chains (Coppi et al. 2007). Because CSP is not cleaved in dermal and ECs with less sulfated HSPGs, sporozoites migrate through these cell types. When they encounter highly sulfated HSPGs of hepatocytes, sporozoites are activated and productively invade these cells (Coppi et al. 2007). Subsequent studies showed that the plasmodium calcium-dependent protein kinase 6 (CDPK-6) is required for the cleavage event, and CDPK-6 deficient sporozoites migrate effectively through other cell types but are unable to productively invade hepatocytes (Coppi et al. 2007). However, how interaction with highly sulfated HSPGs regulates CDPK-6 is not understood.

Review of the complete malaria genome (Carlton et al. 2002; Gardner et al. 2002) has yielded information on several additional proteins involved in hepatocyte invasion. The transmembrane protein TRAP (thrombospondin-related anonymous protein) is present on the sporozoite surface, but its surface expression increases greatly on sporozoites after contact with hepatocytes suggesting a role for TRAP in invasion (Gantt et al. 2000). The extracellular portion of TRAP has two cell-adhesive regions: an integrin-like I-domain (A-domain) and a type 1 thrombospondin motif (TSR) (Menard 2000). Interestingly, TSR binds to HSPGs in the space of Disse (Muller et al. 1993; Robson et al. 1995) and the A-domain binds heparin (McCormick et al. 1999), but the exact role of the TRAP–HSPG interaction in malaria pathogenesis remains to be determined.

Pregnancy-associated malaria is a potentially severe disease leading to maternal anemia and low birth-weight infants because of sequestration of infected erythrocytes in the placenta. A variant of the *P. falciparum* Erythrocyte Membrane Protein 1 (PfEMP1) family named VAR2CSA is expressed only on infected erythrocytes in pregnant women (Salanti et al. 2003) and binds specifically to placental CSPG (Khunrae et al. 2010). In women, high antibody titers against VAR2CSA are protective against low birth-weight babies (Salanti et al. 2004) and VAR2CSA is a candidate antigen for a vaccine against pregnancy-associated malaria (Nielsen et al. 2009).

2.6 Multiple Roles of HSPG in the Pathogenesis of a Single Pathogen

T. gondii is a protozoan parasite able to infect all warm-blooded animals and is estimated to infect one third of humans (Kim and Weiss 2008). Common clinical syndromes associated with *T. gondii* infection include abortion and congenital infection as well as systemic and central nervous system infections in immunocompromised hosts (Kim and Weiss 2008). Infection is transmitted through ingestion of undercooked meats (especially pork) or contact with infected feces from the definitive hosts (felids), although water containing oocysts has been increasingly identified as a source of outbreaks (Kim and Weiss 2008). As an obligate intracellular parasite, *T. gondii* has a complicated life cycle requiring invasion of host cells and formation of a parasitophorus vacuole where the organism can evade host defenses and develop into tachyzoites. Tachyzoites differentiate into bradyzoites, which are contained in thick-walled cysts inside the parasitophorous vacuole and can persist indefinitely – or reactivate if the host becomes immunocompromised (Kim and Weiss 2008).

The first role identified for HSPGs in *T. gondii* infection is its attachment to host cells (Carruthers et al. 2000). Because *T. gondii* can infect a wide range of mammalian hosts and can invade nearly all mammalian and avian cell types, Carruthers et al. (2000) hypothesized that the *T. gondii* receptor is an abundantly expressed and ubiquitous cell surface molecule, such as GAGs (Carruthers et al. 2000). As has been shown for other pathogens, incubation of target cells (human foreskin fibroblasts, in this case) with increasing concentrations of GAGs, including heparin, HS, and CS, inhibited parasite attachment (Carruthers et al. 2000). Parasite attachment to CHO cells with a mutation in xylose transferase, which lacks GAG expression, was significantly decreased supporting a role for HS and CS in attachment (Carruthers et al. 2000). Further, enzymatic removal of HS or CS from the cell surface demonstrated that HS was more important than CS in parasite attachment (Carruthers et al. 2000). Decreased attachment to CHO cells with a defect in NDST-1, an enzyme responsible for *N*-deacetylation/*N*-sulfation of GlcNAc in HS, was seen in vitro (Ortega-Barria and Boothroyd 1999), suggesting the importance of a specific HS modification in attachment. However, the importance of NDST-1 could not be confirmed in a mouse model of disseminated *T. gondii* infection (Bishop et al. 2005), leaving its true contribution uncertain.

Subsequent studies confirmed the requirement for *N*-sulfation in parasite infectivity, but were unable to show that soluble HS could prevent attachment, and demonstrated that tachyzoites were able to attach to and invade $Ndst1^{-/-}$ cells (Bishop et al. 2005). Surprisingly, the effect of NDST-1 deletion on tachyzoites was to decrease the rate of parasite replication in parasitophorous vacuoles (Bishop et al. 2005). A *T. gondii* protein, SAG3, has been identified to bind HSPG and was suggested as the parasite protein, which attaches *T. gondii* to host cells (Jacquet et al. 2001). However, with the role of HSPGs in attachment in doubt, some suggest SAG3 may actually function as an attachment coreceptor by binding to abundant sialic acid residues (Bishop and Esko 2005).

Recent studies into the relationship between *T. gondii* infection and host cell apoptosis have revealed an additional interaction between *T. gondii* and HSPGs. *T. gondii*-infected cells were initially reported to be resistant to apoptosis, which benefits the parasite as it requires a living host cell for replication (Nash et al. 1998). However, several subsequent studies have demonstrated that *T. gondii* infection can induce apoptosis, mostly in neighboring, uninfected host cells (Liesenfeld et al. 1997; Mordue et al. 2001; Nishikawa et al. 2007). Bannai et al. (2008) investigated a *T. gondii* analog to the human protein Programmed Cell Death 5 (TgPDCD5) during apoptosis and during interactions with host cells. TgPDCD5 is a secreted peptide that contains an HSPG-binding motif and localizes mostly to the apical end of the parasite (Bannai et al. 2008). Although unable to induce apoptosis on its own, TgPDCD5 was able to enhance apoptosis in host cells treated with the topoisomerase II inhibitor etoposide and in cooperation with IFN-γ (Bannai et al. 2008). As for human PDCD5, the HSPG-binding motif is required for uptake of TgPDCD5 into host cells (Bannai et al. 2008). The exact role of HSPG binding remains to be fully elucidated, but it is tempting to speculate that HSPG-mediated endocytosis plays a role in TgDCD5 internalization into uninfected host cells where it can then enhance apoptosis.

2.7 HSPG Shedding in Bacterial Pathogenesis

Pathogens are known to enhance the ectodomain shedding of a variety of host cell surface molecules to modulate their environment and enhance virulence (Vollmer et al. 1996; Walev et al. 1996). It has been shown that certain pathogens can induce syndecan-1 shedding either by usurping the host cell machinery or through the pathogen's own sheddase. Several pathogenic organisms have been evaluated for their ability to shed syndecan-1. *Staphylococcus aureus* (Park et al. 2004), *Pseudomonas aeruginosa* (Park et al. 2000), *Streptococcus pneumonia* (Chen et al. 2007b), and *Bacillus anthracis* (Popova et al. 2006) are able to induce shedding. However, several strains of other Gram-positive and Gram-negative bacteria, including *Staphylococcus saprophyticus, Staphylococcus xylosus, Salmonella enteritidis, Salmonella typhimurium,* and *Klebsiella pneumoniae* do not appear to enhance shedding (Park et al. 2000).

2.7.1 Pseudomonas aeruginosa

P. aeruginosa is a Gram-negative bacterium that is associated with infections of the skin, urinary tract, and lung. Particularly concerning is its role as a cause of ventilator-associated pneumonia in hospitalized patients, which complicates the courses of up to 25% of intubated patients, and can carry up to a 10-fold increased risk of mortality (Chastre and Fagon 2002). Culture supernatants of various strains of *P. aeruginosa* were observed to increase syndecan-1 shedding at least fourfold over baseline in multiple cell types (Park et al. 2000). The secreted protein responsible for inducing syndecan-1 shedding was identified as LasA (Kessler et al. 1997), which is a known virulence factor of *P. aeruginosa* in animal models of corneal (Preston et al. 1997) and lung infection (Woods et al. 1982; Blackwood et al. 1983). Although LasA has enzymatic activity (Kessler et al. 1997), it does not hydrolyze syndecan-1. Consistent with these data, studies using PTK inhibitors or metalloproteinase inhibitors (peptide hydroxamate) suggested that LasA induces syndecan-1 shedding via activation of the host cell's shedding machinery (Park et al. 2000).

The physiological role of *P. aeruginosa*-induced syndecan-1 shedding was studied first in a mouse model of pneumonia. Seven-day-old syndecan-1 null mice (*Sdc1*$^{-/-}$) markedly resisted intranasal infection with *P. aeruginosa* as measured by lung and spleen bacterial burden, pneumonia and inflammation as seen on histopathology, as well as overall mortality (Park et al. 2001). As discussed earlier, for the majority of pathogens, HSPGs are used as attachment sites. However, *P. aeruginosa* did not bind to syndecan-1, and excess heparin had no effect on the adhesion of *P. aeruginosa* to lung epithelial cells in vitro (Park et al. 2001). Intranasal inoculation of *Sdc1*$^{-/-}$ mice with heparin or purified syndecan-1 ectodomain at the time of intranasal infection with *P. aeruginosa* restored susceptibility to infection, but inoculation with CS or HS-free syndecan-1 core proteins had no effect (Park et al. 2001). Syndecan-1 ectodomains are detected in the bronchoalveolar lavage fluid after infection with *P. aeruginosa* or after instillation of LasA. Addition of the metalloproteinase inhibitor BB1101 was able to decrease the amount of syndecan-1, but not syndecan-4 ectodomains shed in vivo in response to *P. aeruginosa* infection or Las A instillation, consistent with in vitro results and supporting the specific shedding of syndecan-1 by LasA (Park et al. 2001). Together, these findings suggest that *P. aeruginosa* subverts syndecan-1 shedding to promote its pathogenesis by inhibiting host defense mechanisms in the airspace.

P. aeruginosa is a key pathogen in sepsis following thermal injury (burn wounds). *Sdc1*$^{-/-}$ mice were found to be resistant to *P. aeruginosa* sepsis following thermal injury (Haynes et al. 2005). Absence of syndecan-1 had no effect on the ability of *P. aeruginosa* to colonize burned tissue locally, but led to significantly less systemic spread of infection – and this resistance to systemic spread could be overcome by local injection of HS (Haynes et al. 2005). Thus, syndecan-1 shedding may play a specific role in the dissemination of *P. aeruginosa* in the context of burn infection, but precisely how this is accomplished is incompletely understood.

The ability of *P. aeruginosa* to attach to polarized epithelial cells has been investigated recently. N-glycans, expressed at both the apical and basolateral surface of epithelial monolayers of renal or airway cells, mediate binding and entry of *P. aeruginosa* at the apical surface (Bucior et al. 2010). *P. aeruginosa* binding to the basolateral surface, in contrast, is mediated by HSPGs (Bucior et al. 2010). Tissue injury or epithelial damage increases apical surface expression of N-glycans leading to increased *P. aeruginosa* attachment. Using incompletely polarized cells as a model of injured tissue, it was shown that HSPGs are also upregulated at the apical surface further increasing *P. aeruginosa* binding and subsequent tissue damage (Bucior et al. 2010). The upregulation of expression of both N-glycans and HSPGs in the context of tissue damage and repair may explain the increased propensity for *P. aeruginosa* to cause infections in this context.

2.7.2 *Staphylococcus aureus*

S. aureus is an important Gram-positive bacterial pathogen of humans that causes a variety of diseases ranging from superficial skin and soft tissue infections to serious invasive diseases such as pneumonia, osteomyelitis, endocarditis, and sepsis. *S. aureus* expresses several different classes of virulence factors including pore-forming toxins, such as α-toxin and Panton-Valentine leukocidin, secreted toxins such as β-toxin, toxic shock syndrome toxin-1 (TSST-1) and enterotoxins, cell-wall attached proteins such as protein A, and cell wall components such as peptidoglycan and lipoteichoic acid (Gordon and Lowy 2008).

S. aureus has been shown to induce syndecan-1 shedding through its cytotoxic virulence factors α-toxin and β-toxin. *S. aureus* α-toxin has been established or proposed as a virulence factor in many staphylococcal diseases, such as pneumonia (McElroy et al. 1999; Bartlett et al. 2008), sepsis (Buerke et al. 2002), endocarditis (Bayer et al. 1997), meningitis (Kielian et al. 2001; Maślińska et al. 2004), keratitis (Callegan et al. 1994; Girgis et al. 2005), dermatitis (Ezepchuk et al. 1996), septic arthritis (Nilsson et al. 1999), and mastitis (Bramley et al. 1989). Alpha-toxin is secreted as a soluble monomer and forms heptameric transmembrane pores in target cell membranes (Bhakdi and Tranum-Jensen 1991). Alpha-toxin has additional biological effects on host cells as it binds to a putative glycoprotein receptor (Bhakdi and Tranum-Jensen 1991) and activates intracellular signaling and modulates cellular processes (Bantel et al. 2001; Dragneva et al. 2001; Rose et al. 2002; Park et al. 2004; Haugwitz et al. 2006; Ratner et al. 2006; Liang and Ji 2007). However, the α-toxin receptor has been elusive, and precisely how the noncytolytic activities of α-toxin contribute to *S. aureus* virulence remains to be determined. To date, β1 integrin (Liang and Ji 2007), caveolin-1 (Vijayvargia et al. 2004), and band-3 (Maharaj and Fackrell 1980) have been proposed to be α-toxin receptors, but it remains to be determined if these indeed signal upon α-toxin binding.

S. aureus β-toxin has also been established or proposed as a virulence factor in several staphylococcal diseases, such as keratitis (O'Callaghan et al. 1997) and mastitis (Bramley et al. 1989), although it is one of the least-studied staphylococcal toxins and is found in few clinical isolates (Aarestrup et al. 1999). Beta-toxin is a Mg^{2+}-dependent sphingomyelinase that generates phosphorylcholine and the bioactive secondary messenger ceramide by hydrolyzing host cell membrane sphingomyelin (Doery et al. 1963; Marques et al. 1989; Vollmer et al. 1996). Beta-toxin does not lyse most cell types, but leaves them vulnerable to a number of other lytic agents. In fact, the cytotoxic effect of β-toxin is highly cell type- and species-specific, suggesting that its primary virulence activity is to modulate host processes that affect pathogenesis rather than to directly kill host cells.

Both α-toxin and β-toxin apparently induce syndecan-1 shedding by stimulating the host cell's shedding mechanism. The metalloproteinase inhibitor GM6001 and PTK inhibitor Tyrphostin A25 inhibited toxin-induced syndecan-1 shedding when they were coincubated with toxins and host cells, but not when preincubated with toxins and removed prior to incubation with host cells (Park et al. 2004). These data suggest that both α-toxin and β-toxin activate a similar PTK-dependent, metalloproteinase-mediated shedding mechanism of host cells. The physiological significance of toxin-induced syndecan-1 shedding remains to be established. However, several data suggest that this is an important virulence activity. Instillation of β-toxin-positive (*hlb*+) *S. aureus* or purified β-toxin into mice induced syndecan-1 shedding from type II alveolar epithelial cells and caused pulmonary edema and inflammation, particularly neutrophilic influx into the airways (Hayashida et al. 2009). The capacity of β-toxin to induce syndecan-1 shedding is dependent on its sphingomyelinase activity because mutant proteins lacking this activity failed to trigger shedding and cause lung injury when inoculated intranasally (Hayashida et al. 2009). Consistent with these data, $Sdc1^{-/-}$ mice injected with purified β-toxin showed minimal inflammation, supporting the role of *S. aureus* β-toxin and its sphingomyelinase activity in enhancing syndecan-1 shedding leading to pulmonary inflammation and injury (Hayashida et al. 2009). These results suggest a model in which β-toxin acts as a virulence factor not through direct cytotoxicity, but rather through enhanced neutrophil infiltration which occurs in a syndecan-1 dependent manner (Hayashida et al. 2009) (Fig. 2.2).

The *S. aureus* hetero-oligomeric pore-forming toxin Panton-Valentine leukocidin (PVL) uses HS in a different manner. The signal peptide of the LukS component of PVL is released outside the cytosolic membrane after cleavage by the signal peptidase and the C-terminus can associate with the bacterial cell wall (Tristan et al. 2009). The positively charged N-terminus is then accessible to interact with negatively charged HS chains in the ECM, possibly forming a bridge between bacteria and ECM which explains the increased adherence to damaged epithelial cells seen in PVL+ strains of *S. aureus* (Tristan et al. 2009).

2.7.3 *Streptococcus pneumoniae*

S. pneumoniae is a Gram-positive bacterial pathogen causing diseases such as pneumonia, otitis media, sinusitis, and meningitis. Similar to *S. aureus* and *P. aeruginosa*,

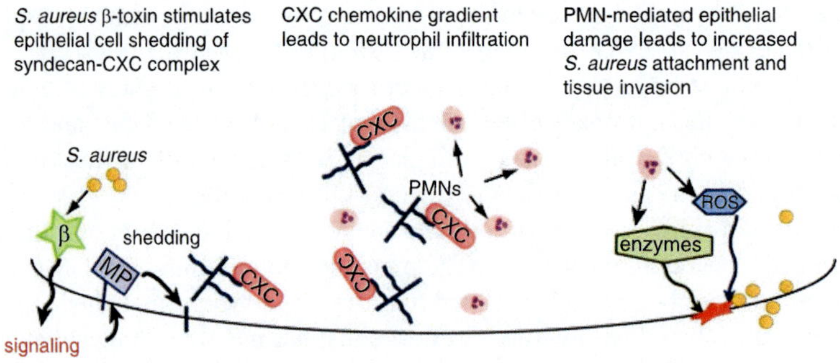

Fig. 2.2 Mechanisms of microbial subversion of HSPG shedding. *S. aureus* β-toxin induces shedding of syndecan-1, which is shown here bound to the cytokine IL-8. *S. aureus* β-toxin stimulates an intracellular signaling pathway through its sphingomyelinase activity, which leads to acceleration of the metalloproteinase-mediated cleavage of syndecan-1 ectodomain-IL-8 complex at the cell surface. Shedding of the syndecan-1-IL-8 complex creates a chemotactic gradient guiding neutrophils to migrate into the airspace leading to inflammation and tissue damage

S. pneumoniae induces syndecan-1 shedding from the surface of host cells in culture. However, available data indicate that *S. pneumoniae* directly sheds syndecan-1 ectodomains (Chen et al. 2007b). As such, the size of syndecan-1 ectodomains shed by *S. pneumoniae* is smaller than ectodomains produced by constitutive shedding. More importantly, ZmpC, one of the pneumococcal zinc metalloproteinases, has been shown to possess syndecan-1 sheddase activity in vitro (Chen et al. 2007b). These observations suggest that several major opportunistic bacterial pathogens use distinct mechanisms to induce syndecan-1 shedding, which can enhance bacterial virulence in vivo. However, it remains to be established that induction of syndecan-1 shedding is an important virulence activity of *S. pneumoniae*.

2.8 Future Perspectives

Studies during the last several decades have demonstrated that viral, bacterial, and parasitic pathogens elaborate factors that interact with host HSPGs. Some pathogens also express factors that modulate the expression pattern of HSPGs by inducing shedding. However, it is still not clear if these interactions benefit the host or the bacteria. Several in vivo studies suggest that induction of syndecan-1 shedding is a critical virulence activity of certain bacterial pathogens. However, the fact that the majority of intracellular pathogens exploit cell surface HSPGs for its attachment and cellular invasion also suggest that syndecan shedding is a host defense mechanism that rapidly and transiently downregulates microbial attachment sites. Perhaps only highly effective pathogens, such as *S. aureus*, have adapted or evolved to subvert this innate host mechanism. Future studies should be directed at testing the physiological significance of HSPG–pathogen interaction using animal models of infection that simulate human disease. In vivo studies using mice deficient in specific HS biosynthetic enzymes and mice deficient in certain HSPGs or HS enzymes in a cell-specific manner should also be attempted to define the molecular and cellular features of the HSPG–pathogen interaction. These studies should further define the underlying mechanisms and the physiological relevance of HSPG–pathogen interactions.

Although improvements in hygienic, prophylactic, and therapeutic interventions have significantly reduced the incidence of infections in the last century, infectious diseases continue to be a major public health threat. Further, the continuous emergence of drug-resistant strains is adding to this threat. Because available data suggest that the capacity to subvert HSPGs to promote infection is a pathogenic mechanism used by many pathogens, the HSPG–pathogen interaction is a potential candidate for novel antimicrobial therapy against a broad range of infectious diseases. For example, de Witte et al. (2008) showed that papillomavirus VLPs interact with HSPGs on the surface of DCs, as binding can be prevented by treatment of DCs with heparinase II. As DCs express syndecan-3 and play a critical role in immunity against HPVs, the HS moiety of syndecan-3 may be a target for anti-HPV therapy. Similarly, periodate-cleaved heparin fragments that lack antithrombin-III binding activity and thus lack anticoagulant properties have been shown to retain several key functions that may have therapeutic benefits in the treatment of severe malaria; they can block merozoite invasion of erythrocytes, disrupt rosettes, inhibit endothelial binding of *Plasmodium* in vitro and reverse sequestration in vivo in rat and macaque models of malaria (Vogt et al. 2006). These data suggest that low molecular weight HS/heparin compounds that do not cause the unwanted side effects of heparin therapy, such as bleeding and thrombocytopenia, are potential candidates for antimalarial therapy. Deciphering the key mechanisms of HSPG–pathogen interactions should provide important insights into the design and development of novel HS-based antimicrobial strategies.

References

Aarestrup F, Larsen H et al (1999) Frequency of alpha- and beta-haemolysin in *Staphylococcus aureus* of bovine and human origin. A comparison between pheno- and genotype and variation in phenotypic expression. APMIS 107:425–430

Akula S, Pramod N et al (2001) Human herpesvirus 8 envelope-associated glycoprotein B interacts with heparan sulfate-like moieties. Virology 284:235–249

Albini A, Soldi R et al (1996) The angiogenesis induced by HIV-1 tat protein is mediated by the Flk-1/KDR receptor on vascular endothelial cells. Nat Med 2:1371–1375

Albini A, Ferrini S et al (1998) HIV-1 Tat protein mimicry of chemokines. Proc Natl Acad Sci USA 95:13153–13158

Alexopoulou A, Multhaupt H et al (2007) Syndecans in wound healing, inflammation and vascular biology. Int J Biochem Cell Biol 39:505–528

Alfsen A, Yu H et al (2005) HIV-1-infected blood mononuclear cells form an integrin- and agrin-dependent viral synapse to induce efficient HIV-1 transcytosis across epithelial cell monolayer. Mol Biol Cell 16:4267–4279

Alvarez-Dominguez C, Carrasco-Marin E et al (1993) Role of complement component C1q in phagocytosis of Listeria monocytogenes by murine macrophage-like cell lines. Infect Immun 61:3664–3672

Alvarez-Dominguez C, Vazquez-Boland J-A et al (1997) Host cell Heparan sulfate proteoglycans mediate attachment and entry of *Listeria monocytogenes*, and the listerial surface protein ActA is involved in heparan sulfate receptor recognition. Infect Immun 65(1):78–88

Andrian E, Grenier D et al (2005) *Porphyromonas gingivalis* lipopolysaccharide induces shedding of syndecan-1 expressed by gingival epithelial cells. J Cell Physiol 204:178–183

Andrian E, Grenier D et al (2006) *Porphyromonas gingivalis* gingipains mediate the shedding of syndecan-1 from the surface of gingival epithelial cells. Oral Microbiol Immunol 21:123–128

Argyris E, Kulkosky J et al (2004) The perlecan heparan sulfate proteoglycan mediates cellular uptake of HIV-1 Tat through a pathway responsible for biological activity. Virology 330: 481–486

Arungundram S, Al-Mafraji K et al (2009) Modular synthesis of heparan sulfate oligosaccharides for structure-activity relationship studies. J Am Chem Soc 131(47):17394–17405

Baeuerle P, Huttner W (1986) Chlorate – a potent inhibitor of protein sulfation in intact cells. Biochem Biophys Res Commun 141:870–877

Baleux F, Loureiro-Morais L et al (2009) A synthetic CD4-heparan sulfate glycoconjugate inhibits CCR5 and CXCR4 HIV-1 attachment and entry. Nat Chem Biol 5(10):743–748

Banerjee M, Copp J et al (2004) GW domains of the Listeria monocytogenes invasion protein InlB are required for potentiation of Met activation. Mol Microbiol 52:257–271

Bannai H, Nishikawa Y et al (2008) Programmed cell death 5 from *Toxoplasma gondii*: a secreted molecule that exerts a pro-apoptotic effect on host cells. Mol Biochem Parasitol 159:112–120

Bantel H, Sinha B et al (2001) alpha-Toxin is a mediator of *Staphylococcus aureus*-induced cell death and activates caspases via the intrinsic death pathway independently of death receptor signaling. J Cell Biol 155:637–648

Baron M, Bolduc G et al (2004) Alpha C protein of group B Streptococcus binds host cell surface glycosaminoglycan and enters cells by an actin-dependent mechanism. J Biol Chem 279:24714–24723

Barth H, Schnober E et al (2006) Viral and cellular determinants of the hepatitis C virus envelope-heparan sulfate interaction. J Virol 80:10579–10590

Bartlett A, Foster T et al (2008) Alpha-toxin facilitates the generation of CXC chemokine gradients and stimulates neutrophil homing in *Staphylococcus aureus* pneumonia. J Infect Dis 198:1529–1535

Bayer A, Ramos M et al (1997) Hyperproduction of alpha-toxin by *Staphylococcus aureus* results in paradoxically reduced virulence in experimental endocarditis: a host defense role for platelet microbicidal proteins. Infect Immun 65:4652–4660

Bernfield M, Gotte M et al (1999) Functions of cell surface heparan sulfate proteoglycans. Annu Rev Biochem 68:729–777

Bhakdi S, Tranum-Jensen J (1991) Alpha-toxin of *Staphylococcus aureus*. Microbiol Rev 55:733–751

Bishop J, Esko J (2005) The elusive role of heparan sulfate in *Toxoplasma gondii* infection. Mol Biochem Parasitol 139:267–269

Bishop J, Crawford B et al (2005) Cell surface heparan sulfate promotes replication of Toxoplasma gondii. Infect Immun 73:5395–5401

Blackwood L, Stone R et al (1983) Evaluation of *Pseudomonas aeruginosa* exotoxin A and elastase as virulence factors in acute lung infection. Infect Immun 39:198–201

Bonazzi M, Veiga E et al (2008) Successive post-translational modifications of E-cadherin are required for InlA-mediated internalization of Listeria monocytogenes. Cell Microbiol 10:2208–2222

Bossis I, Roden R et al (2005) Interaction of tSNARE syntaxin 18 with the papillomavirus minor capsid protein mediates infection. J Virol 79:6723–6731

Boyd A, Sory M et al (1998) Heparin interferes with translocation of Yop proteins into HeLa cells and binds to LcrG, a regulatory component of the Yersinia Yop apparatus. Mol Microbiol 27:425–436

Boyle K, Compton T (1998) Receptor-binding properties of a soluble form of human cytomegalovirus glycoprotein B. J Virol 72:1826–1833

Bramley A, Patel A et al (1989) Roles of alpha-toxin and beta-toxin in virulence of *Staphylococcus aureus* for the mouse mammary gland. Infect Immun 57:2489–2494

Braun L, Ghebrehiwet B et al (2000) gC1q-R/p32, a C1q-binding protein, is a receptor for the InlB invasion protein of Listeria monocytogenes. EMBO J 19:1458–1466

Brule S, Charnaux N et al (2006) The shedding of syndecan-4 and syndecan-1 from HeLa cells and human primary macrophages is accelerated by SDF-1/CXCL12 and mediated by the matrix metalloproteinase-9. Glycobiology 16:488–501

Bucior I, Mostov K et al (2010) Pseudomonas aeruginosa-mediated damage requires distinct receptors at the apical and basolateral surfaces of the polarized epithelium. Infect Immun 78 (3):939–953

Buck C, Thompson C et al (2006) Carrageenan is a potent inhibitor of papillomavirus infection. PLoS Pathog 2:e69

Buerke M, Sibelius U et al (2002) *Staphylococcus aureus* alpha toxin mediates polymorphonuclear leukocyte-induced vasocontraction and endothelial dysfunction. Shock 17:30–35

Bugatti A, Urbinati C et al (2007) Heparin-mimicking sulfonic acid polymers as multitarget inhibitors of human immunodeficiency virus type 1 Tat and gp120 proteins. Antimicrob Agents Chemother 51:2337–2345

Callegan M, Engel L et al (1994) Corneal virulence of *Staphylococcus aureus*: roles of alpha-toxin and protein A in pathogenesis. Infect Immun 62:2478–2482

Capurro M, Wanless I et al (2003) Glypican-3: a novel serum and histochemical marker for hepatocellular carcinoma. Gastroenterology 125:89–97

Carlton J, Angiuoli S et al (2002) Genome sequence and comparative analysis of the model rodent malaria parasite Plasmodium yoelii yoelii. Nature 519:512–519

Carruthers V, Håkansson S et al (2000) *Toxoplasma gondii* uses sulfated proteoglycans for substrate and host cell attachment. Infect Immun 68:4005–4011

Ceballos A, Lenicov F et al (2009) Spermatozoa capture HIV-1 through heparan sulfate and efficiently transmit the virus to dendritic cells. J Exp Med 206(12):2717–2733

Chang H, Samaniego F et al (1997) HIV-1 Tat protein exits from cells via a leaderless secretory pathway and binds to extracellular matrix-associated heparan sulfate proteoglycans through its basic region. AIDS 11(12):1421–1431

Chastre J, Fagon J (2002) Ventilator-associated Pneumonia Jean Chastre and Jean-Yves Fagon. Am J Respir Crit Care Med 165:867–903

Chavez-Salinas S, Ceballos-Olvera I et al (2008) Heat shock effect upon dengue virus replication into U937 cells. Virus Res 128:111–118

Chen R, Lander A (2001) Mechanisms underlying preferential assembly of heparan sulfate on glypican. J Biol Chem 276:7507–7517

Chen J, Stephens R (1997) Chlamydia trachomatis glycosaminoglycan-dependent and independent attachment to eukaryotic cells. Microb Pathog 22(1):23–30

Chen Y, Maguire T et al (1997) Dengue virus infectivity depends on envelope protein binding to target cell heparan sulfate. Nat Med 3:866–871

Chen J, Jones C et al (2007a) Using an enzymatic combinatorial approach to identify anticoagulant heparan sulfate structures. Chem Biol 14:986–993

Chen Y, Hayashida A et al (2007b) Streptococcus pneumoniae sheds Syndecan-1 ECTODOMAINS through ZmpC, a metalloproteinase virulence factor. J Biol Chem 282:159–167

Chen Y, Götte M et al (2008) Microbial subversion of heparan sulfate proteoglycans. Mol Cells 26 (5):415–426

Cheshenko N, Liu W et al (2007) Multiple receptor interactions trigger release of membrane and intracellular calcium stores critical for herpes simplex virus entry. Mol Biol Cell 18: 3119–3130

Chirivi R, Taraboletti G et al (1999) Human immunodeficiency virus-1 (HIV-1)-Tat protein promotes migration of acquired immunodeficiency syndrome-related lymphoma cells and enhances their adhesion to endothelial cells. Blood 94:1747–1754

Christensen N, Cladel N et al (1995) Postattachment neutralization of papillomaviruses by monoclonal and polyclonal antibodies. Virology 207:136–142

Chung M, Popova T et al (2006) Secreted neutral metalloproteases of *Bacillus anthracis* as candidate pathogenic factors. J Biol Chem 281:31408–31418

Clapham P, McKnight A (2001) HIV-1 receptors and cell tropism. Br Med Bull 58:43–59

Cohen A, Wood D et al (1998) Human CASK/LIN-2 binds Syndecan-2 and Protein 4.1 and localizes to the basolateral membrane of epithelial cells. J Cell Biol 142:129–138

Coppi A, Pinzon-Ortiz C et al (2005) The Plasmodium circumsporozoite protein is proteolytically processed during cell invasion. J Exp Med 201(1):27–33

Coppi A, Tewari R et al (2007) Heparan sulfate proteoglycans provide a signal to Plasmodium sporozoites to stop migrating and productively invade cells. Cell Host Microbe 2(5):316–327

Crim R, Audet S et al (2007) Identification of linear heparin-binding peptides derived from human respiratory syncytial virus fusion glycoprotein that inhibit infectivity. J Virol 81:261–271

Crublet E, Andrieu J et al (2008) The HIV-1 envelope glycoprotein gp120 features four heparan sulfate binding domains, including the co-receptor binding site. J Biol Chem 283(22): 15193–15200

David G, Lorie V et al (1990) Molecular cloning of a phosphatidylinositol-anchored membrane heparan sulfate proteoglycan from human lung fibroblasts. J Cell Biol 111:3165–3176

Day P, Gambhira R et al (2008) Mechanisms of human papillomavirus type 16 neutralization by 12 cross-neutralizing and 11 type-specific antibodies. J Virol 82(9):4638–4646
de Haan C, Haijema B et al (2008) Cleavage of group 1 coronavirus spike proteins: how furin cleavage is traded off against heparan sulfate binding upon cell culture adaptation. J Virol 82:6078–6083
de Vries F, Cole R et al (2002) *Neisseria meningitidis* producing the Opc adhesin binds epithelial cell proteoglycan receptors. Mol Microbiol 27:1203–1212
de Witte L, Bobardt M et al (2007) Syndecan-3 is a dendritic cell-specific attachment receptor for HIV-1. Proc Natl Acad Sci USA 104:19464–19469
de Witte L, Zoughlami Y et al (2008) Binding of human papilloma virus L1 virus-like particles to dendritic cells is mediated through heparan sulfates and induces immune activation. Immunobiology 212:679–691
Dewhurst S, Gelbard H et al (1996) Neuropathogenesis of AIDS. Mol Med Today 2(1):16–23
Ding K, Lopez-Burks M et al (2005) Growth factor–induced shedding of syndecan-1 confers glypican-1 dependence on mitogenic responses of cancer cells. J Cell Biol 171:729–738
Doery H, Magnusson B et al (1963) A phospholipase in staphylococcal toxin which hydrolyses sphingomyelin. Nature 198:1091–1092
Dragneva Y, Anuradha C et al (2001) Subcytocidal attack by staphylococcal alpha-toxin activates NF-kappaB and induces interleukin-8 production. Infect Immun 69:2630–2635
Dramsi S, Biswas I et al (1995) Entry of Listeria monocytogenes into hepatocytes requires the expression of InlB, a surface protein of the internalin multigene family. Mol Microbiol 16:251–261
Dramsi S, Dehoux P et al (1997) Identification of four new members of the internalin multigene family of Listeria monocytogenes EGD. Infect Immun 65:1615–1625
Drevets D, Campbell P (1991) Roles of complement receptor type 3 in phagocytosis of Listeria monocytogenes by inflammatory mouse peritoneal macrophages. Infect Immun 59:2645–2652
Endo K, Takino T et al (2003) Cleavage of Syndecan-1 by membrane type matrix metalloproteinase-1 stimulates cell migration. J Biol Chem 278:40764–40770
Esko J, Montgomery R (1995) Synthetic glycosides as primers of oligosaccharide biosynthesis and inhibitors of glycoprotein and proteoglycan assembly. In: Ausubel F, Brent R, Kingston R et al. (eds) Current protocols in molecular biology. Greene and Wiley-Interscience, New York, pp 17.11.11–17.11.16
Ezepchuk Y, Leung D et al (1996) Staphylococcal toxins and protein A differentially induce cytotoxicity and release of tumor necrosis factor-alpha from human keratinocytes. J Invest Dermatol 107:603–609
Fears C, Woods A (2006) The role of syndecans in disease and wound healing. Matrix Biol 25:443–456
Fitzgerald M, Wang Z et al (2000) Shedding of Syndecan-1 and -4 ectodomains is regulated by multiple signaling pathways and mediated by a TIMP-3-sensitive metalloproteinase. J Cell Biol 148:811–824
Forsberg E, Pejler G et al (1999) Abnormal mast cells in mice deficient in a heparin-synthesizing enzyme. Nature 400:773–776
Freissler E, Meyer auf der Heyde A et al (2000) Syndecan-1 and syndecan-4 can mediate the invasion of OpaHSPG-expressing *Neisseria gonorrhoeae* into epithelial cells. Cell Microbiol 2:69–82
Frevert U, Sinnis P et al (1993) Malaria circumsporozoite protein binds to heparan sulfate proteoglycans associated with the surface membrane of hepatocytes. J Exp Med 177: 1287–1298
Frick I, Schmidtchen A et al (2003) Interactions between M proteins of *Streptococcus pyogenes* and glycosaminoglycans promote bacterial adhesion to host cells. Eur J Biochem 270: 2303–2311
Fry E, Lea S et al (1999) The structure and function of a foot-and-mouth disease virus-oligosaccharide receptor complex. EMBO J 18:543–554

Fuki I, Kuhn K et al (1997) The syndecan family of proteoglycans. Novel receptors mediating internalization of atherogenic lipoproteins in vitro. J Clin Invest 100:1611–1622

Gantt S, Persson C et al (2000) Antibodies against thrombospondin-related anonymous protein do not inhibit Plasmodium sporozoite infectivity in vivo. Infect Immun 68:3667–3673

Gardner M, Hall N et al (2002) Genome sequence of the human malaria parasite Plasmodium falciparum. Nature 419:498–511

Gatignol A, Jeang T (2000) Tat as a transcriptional activator and a potential therapeutic target for HIV-1. Adv Pharmacol 48:209–227

Germi R, Crance JM et al (2002) Heparan sulfate-mediated binding of infectious Dengue Virus Type 2 and Yellow Fever Virus. Virology 292:162–168

Girgis D, Sloop G et al (2005) Effects of toxin production in a murine model of *Staphylococcus aureus* keratitis. Invest Ophthalmol Vis Sci 46:2064–2070

Giroglou T, Florin L et al (2001) Human papillomavirus infection requires cell surface heparan sulfate. J Virol 75:1565–1570

Gordon R, Lowy F (2008) Pathogenesis of methicillin-resistant *Staphylococcus aureus* infection. Clin Infect Dis 46:S350–S359

Grobe K, Inatani M et al (2005) Cerebral hypoplasia and craniofacial defects in mice lacking heparan sulfate Ndst1 gene function. Development 132:3777–3786

Grootjans J, Zimmermann P et al (1997) Syntenin, a PDZ protein that binds syndecan cytoplasmic domains. Proc Natl Acad Sci USA 94:13683–13688

Hannah J, Menozzi F et al (1994) Sulfated glycoconjugate receptors for the *Bordetella pertussis* adhesin filamentous hemagglutinin (FHA) and mapping of the heparin-binding domain on FHA. Infect Immun 62:5010–5019

Haugwitz U, Bobkiewicz W et al (2006) Pore-forming *Staphylococcus aureus* alpha-toxin triggers epidermal growth factor receptor-dependent proliferation. Cell Microbiol 8:1591–1600

Hayashida K, Johnston D et al (2006) Syndecan-1 expression in epithelial cells is induced by transforming growth factor-beta through a PKA-dependent pathway. J Biol Chem 281: 24365–24374

Hayashida K, Stahl P et al (2008) Syndecan-1 ectodomain shedding is regulated by the small GTPase Rab5. J Biol Chem 283:35435–35444

Hayashida A, Bartlett A et al (2009) Staphylococcus aureus beta-toxin induces lung injury through syndecan-1. Am J Pathol 174:509–518

Haynes A III, Ruda F et al (2005) Syndecan 1 shedding contributes to *Pseudomonas aeruginosa* sepsis. Infect Immun 73:7914–7921

Heino P, Skyldberg B et al (1995) Human papillomavirus type 16 capsids expose multiple type-restricted and type-common antigenic epitopes. J Gen Virol 76:1141–1153

Henry-Stanley M, Hess D et al (2003) Role of heparan sulfate in interactions of *Listeria monocytogenes* with enterocytes. Med Microbiol Immunol 192(2):107–115

Hilgard P, Stockert R (2000) Heparan sulfate proteoglycans initiate dengue virus infection of hepatocytes. Hepatology 32(5):1069–1077

Hippo Y, Watanabe K et al (2004) Identification of soluble NH2-terminal fragment of glypican-3 as a serological marker for early-stage hepatocellular carcinoma. Cancer Res 64:2418–2423

Horonchik L, Tzaban S et al (2005) Heparan sulfate is a cellular receptor for purified infectious prions. J Biol Chem 280:17062–17067

Hsiao J, Chung C et al (1998) Cell surface proteoglycans are necessary for A27L protein-mediated cell fusion: identification of the N-terminal region of A27L protein as the glycosaminoglycan-binding domain. J Virol 72:8374–8379

Hsueh Y, Yang F et al (1998) Direct interaction of CASK/LIN-2 and syndecan heparan sulfate proteoglycan and their overlapping distribution in neuronal synapses. J Cell Biol 142:139–151

Humphries D, Silbert J (1988) Chlorate: a reversible inhibitor of proteoglycan sulfation. Biochem Biophys Res Commun 154:365–371

Hung SL, Lee PL et al (1999) Analysis of the steps involved in dengue virus entry into host cells. Virology 257:156–167

Iozzo R (2005) Basement membrane proteoglycans: from cellar to ceiling. Nat Rev Mol Cell Biol 6:646–656

Isaacs R (1994) *Borrelia burgdorferi* bind to epithelial cell proteoglycans. J Clin Invest 93: 809–819

Jacquet A, Coulon L et al (2001) The surface antigen SAG3 mediates the attachment of *Toxoplasma gondii* to cell-surface proteoglycans. Mol Biochem Parasitol 116:35–44

Jindadamrongwech S, Thepparit C et al (2004) Identification of GRP 78 (BiP) as a liver cell expressed receptor element for dengue virus serotype 2. Arch Virol 149:915–927

Johnson K, Kines R et al (2009) The role of heparan sulfate in HPV attachment and infection of the murine female genital tract. J Virol 83(5):2067–2074

Jonquières R, Pizarro-Cerdá J et al (2001) Synergy between the N- and C-terminal domains of InlB for efficient invasion of non-phagocytic cells by Listeria monocytogenes. Mol Microbiol 42:955–965

Joyce J, Tung J et al (1999) The L1 major capsid protein of human papillomavirus type 11 recombinant virus-like particles interacts with heparin and cell-surface glycosaminoglycans on human keratinocytes. J Biol Chem 274:5810–5822

Kainulainen V, Wang H et al (1998) Syndecans, heparan sulfate proteoglycans, maintain the proteolytic balance of acute wound fluids. J Biol Chem 273:11563–11569

Kato M, Wang H et al (1998) Physiological degradation converts the soluble syndecan-1 ectodomain from an inhibitor to a potent activator of FGF-2. Nat Med 4:691–697

Kawana K, Matsumoto K et al (1998) A surface immunodeterminant of human papillomavirus type 16 minor capsid protein L2. Virology 245:353–359

Kawana K, Yoshikawa H et al (1999) Common neutralization epitope in minor capsid protein L2 of human papillomavirus types 16 and 6. J Virol 73:6188–6190

Kessler E, Safrin M et al (1997) Inhibitors and specificity of *Pseudomonas aeruginosa* LasA. J Biol Chem 272:9884–9889

Khunrae P, Dahlbäck M et al (2010) Full-length recombinant Plasmodium falciparum VAR2CSA binds specifically to CSPG and induces potent parasite adhesion-blocking antibodies. J Mol Biol 397(3):826–834

Kielian T, Cheung A et al (2001) Diminished virulence of an alpha-toxin mutant of *Staphylococcus aureus* in experimental brain abscesses. Infect Immun 69:6902–6911

Kim K, Weiss L (2008) Toxoplasma: the next 100 years. Microb Infect 10:978–984

Kim H, Choi M et al (2004) Role of Syndecan-4 in the cellular invasion of *Orientia tsutsugamushi*. Microb Pathog 36:219–225

Kines R, Thompson C et al (2009) The initial steps leading to papillomavirus infection occur on the basement membrane prior to cell surface binding. Proc Natl Acad Sci USA 106(48): 20458–20463

Kinnunen T, Kaksonen M et al (1998) Cortactin-Src kinase signaling pathway is involved in N-syndecan-dependent neurite outgrowth. J Biol Chem 273:10702–10708

Kirnbauer R, Booy F et al (1992) Papillomavirus L1 major capsid protein self-assembles into virus-like particles that are highly immunogenic. Proc Natl Acad Sci USA 89:12180–12184

Klasse P, Moore J (2004) Is there enough gp120 in the body fluids of HIV-1-infected individuals to have biologically significant effects? Virology 323:1–8

Kobayashi K, Kato K et al (2010) Plasmodium falciparum BAEBL binds to heparan sulfate proteoglycans on the human erythrocyte surface. J Biol Chem 285(3):1716–1725

Kreugera J, Pereza L et al (2004) Opposing activities of dally-like glypican at high and low levels of wingless morphogen activity. Dev Cell 7:503–512

Lasa I, Cossart P (1996) Actin-based bacterial motility: towards a definition of the minimal requirements. Trends Cell Biol 6:109–114

Lee E, Hall R et al (2004) Common E protein determinants for attenuation of glycosaminoglycan-binding variants of Japanese encephalitis and West Nile viruses. J Virol 78:8271–8280

Li Q, Park P et al (2002) Matrilysin shedding of syndecan-1 regulates chemokine mobilization and transepithelial efflux of neutrophils in acute lung injury. Cell 111:635–646

Liang X, Ji Y (2007) Involvement of alpha5beta1-integrin and TNF-alpha in *Staphylococcus aureus* alpha-toxin-induced death of epithelial cells. Cell Microbiol 9:1809–1821

Liesenfeld O, Kosek J et al (1997) Gamma interferon induces Fas-dependent apoptosis of Peyer's patch T cells in mice following peroral infection with *Toxoplasma gondii*. Infect Immun 65:4682–4689

Lima A, Almeida P et al (2002) Heparan sulfate modulates kinin release by *Trypanosoma cruzi* through the activity of cruzipain. J Biol Chem 277:5875–5881

Lindblom A, Fransson L (1990) Endothelial heparan sulphate: compositional analysis and comparison of chains from different proteoglycan populations. Glycoconj J 7(6):545–562

Lingnau A, Domann E et al (1995) Expression of Listeria monocytogenes EGD inlA and inlB genes, whose products mediate bacterial entry into tissue culture cell lines, by PrfA-dependent and -independent mechanisms. Infect Immun 64:1002–1006

Liu W, Gissmann L et al (1997) Sequence close to the N-terminus of L2 protein is displayed on the surface of bovine papillomavirus type 1 virions. Virology 227:474–483

Love D, Esko J et al (1993) A heparin-binding activity on *Leishmania* amastigotes which mediates adhesion to cellular proteoglycans. J Cell Biol 123:759–766

Lyon M, Deakin J et al (1994) Liver heparan sulfate structure. A novel molecular design. J Biol Chem 269(15):11208–11215

Maharaj I, Fackrell H (1980) Rabbit erythrocyte band 3: a receptor for staphylococcal alpha toxin. Can J Microbiol 26:524–527

Marques M, Weller P et al (1989) Phosphatidylinositol-specific phospholipase C, a possible virulence factor of Staphylococcus aureus. J Clin Microbiol 27:2451–2454

Maślińska D, Laure-Kamionowska M et al (2004) Toll-like receptors in rat brains injured by hypoxic-ischaemia or exposed to staphylococcal alpha-toxin. Folia Neuropathol 42:125–132

McCormick C, Tuckwell D et al (1999) Identification of heparin as a ligand for the A-domain of Plasmodium falciparum thrombospondin-related adhesion protein. Mol Biochem Parasitol 100:111–124

McElroy M, Harty H et al (1999) Alpha-toxin damages the air-blood barrier of the lung in a rat model of *Staphylococcus aureus*-induced pneumonia. Infect Immun 67:5541–5544

Menard R (2000) The journey of the malaria sporozoite through its hosts: two parasite proteins lead the way. Microb Infect 2:633–642

Mengaud J, Ohayon H et al (1996) E-cadherin is the receptor for internalin, a surface protein required for entry of L. monocytogenes into epithelial cells. Cell 84:923–932

Mertens G, Van der Schueren B et al (1996) Heparan sulfate expression in polarized epithelial cells: the apical sorting of glypican (GPI-anchored proteoglycan) is inversely related to its heparan sulfate content. J Cell Biol 132:487–497

Moelleken K, Hegemann J (2008) The Chlamydia outer membrane protein OmcB is required for adhesion and exhibits biovar-specific differences in glycosaminoglycan binding. Mol Microbiol 67:403–419

Mordue D, Monroy F et al (2001) Acute toxoplasmosis leads to lethal overproduction of Th1 cytokines. J Immunol 167:4574–4584

Muller H, Reckman I et al (1993) Thrombospondin related anonymous protein (TRAP) of Plasmodium falciparum binds specifically to sulfated glycoconjugates and to HepG2 hepatoma cells suggesting a role for this molecule in sporozoite invasion of hepatocytes. EMBO J 12: 2881–2889

Murthy K, Smith S et al (2001) Crystal structure of a complement control protein that regulates both pathways of complement activation and binds heparan sulfate proteoglycans. Cell 104:301–311

Naguleswaran A, Cannas A et al (2002) Vero cell surface proteoglycan interaction with the microneme protein NcMIC(3) mediates adhesion of *Neospora caninum* tachyzoites to host cells unlike that in *Toxoplasma gondii*. Int J Parasitol 32:695–704

Nakatsura T, Kageshita T et al (2004) Identification of Glypican-3 as a novel tumor marker for melanoma. Clin Cancer Res 10:6612–6621

Nash P, Purner M et al (1998) *Toxoplasma gondii*-infected cells are resistant to multiple inducers of apoptosis. J Immunol 160:1824–1830

Nickells J, Cannella M et al (2008) Neuroadapted yellow fever virus strain 17D: a charged locus in domain III of the E protein governs heparin binding activity and neuroinvasiveness in the SCID mouse model. J Virol 82(24):12510–12519

Nielsen M, Pinto V et al (2009) Induction of adhesion-inhibitory antibodies against placental Plasmodium falciparum parasites by using single domains of VAR2CSA. Infect Immun 77:2482–2487

Nikolova V, Koo C-Y et al (2009) Differential roles for membrane-bound and soluble syndecan-1 (CD138) in breast cancer progression. Carcinogenesis 30:397–407

Nilsson I, Hartford O et al (1999) Alpha-toxin and gamma-toxin jointly promote *Staphylococcus aureus* virulence in murine septic arthritis. Infect Immun 67:1045–1049

Nishikawa Y, Kawase O et al (2007) *Toxoplasma gondii* infection induces apoptosis in noninfected macrophages: role of nitric oxide and other soluble factors. Parasite Immunol 29:375–385

Noel G, Love D et al (1994) High-molecular-weight proteins of nontypeable *Haemophilus influenzae* mediate bacterial adhesion to cellular proteoglycans. Infect Immun 62:4028–4033

Noonan D, Albini A (2000) From the outside in: extracellular activities of HIV Tat. Adv Pharmacol 48:229–250

O'Callaghan R, Callegan M et al (1997) Specific roles of alpha-toxin and beta-toxin during *Staphylococcus aureus* corneal infection. Infect Immun 65:1571–1578

Oliveira FJ, Alves C et al (2008) *Trypanosoma cruzi* heparin-binding proteins and the nature of the host cell heparan sulfate-binding domain. Microb Pathog 44:329–338

Opie S, Warrington KJ et al (2003) Identification of amino acid residues in the capsid proteins of adeno-associated virus type 2 that contribute to heparan sulfate proteoglycan binding. J Virol 77:6995–7006

Ortega-Barria E, Boothroyd J (1999) A Toxoplasma lectin-like activity specific for sulfated polysaccharides is involved in host cell infection. J Biol Chem 274:1267–1276

Pallerla S, Lawrence R et al (2008) Altered heparan sulfate structure in mice with deleted NDST3 gene function. J Biol Chem 283:16885–16894

Park P, Pier G et al (2000) Syndecan-1 shedding is enhanced by LasA, a secreted virulence factor of *Pseudomonas aeruginosa*. J Biol Chem 275:3057–3064

Park P, Pier G et al (2001) Exploitation of syndecan-1 shedding by *Pseudomonas aeruginosa* enhances virulence. Nature 411:98–102

Park P, Foster T et al (2004) Activation of Syndecan-1 ectodomain shedding by Staphylococcus aureus α-toxin and β-toxin. J Biol Chem 279:251–258

Patterson N, Smith J et al (2005) Human papillomavirus type 31b infection of human keratinocytes does not require heparan sulfate. J Virol 79:6838–6847

Pethe K, Alonso S et al (2001) The heparin-binding haemagglutinin of *M. tuberculosis* is required for extrapulmonary dissemination. Nature 412:190–194

Piñon J, Klasse P et al (2003) Human T-cell leukemia virus type 1 envelope glycoprotein gp46 interacts with cell surface heparan sulfate proteoglycans. J Virol 77:9922–9930

Poon G, Gariepy J (2007) Cell-surface proteoglycans as molecular portals for cationic peptide and polymer entry into cells. Biochem Soc Trans 35:788–793

Popova T, Millis B et al (2006) Acceleration of epithelial cell syndecan-1 shedding by anthrax hemolytic virulence factors. BMC Microbiol 6:8

Portnoy D, Chakraborty T et al (1992) Molecular determinants of Listeria monocytogenes pathogenesis. Infect Immun 60:1263–1267

Preston M, Seed P et al (1997) Contribution of proteases and LasR to the virulence of *Pseudomonas aeruginosa* during corneal infections. Infect Immun 65:3086–3090

Pruessmeyer J, Martin C et al (2010) A disintegrin and metalloproteinase 17 (ADAM17) mediates inflammation-induced shedding of syndecan-1 and -4 by lung epithelial cells. J Biol Chem 285 (1):555–564

Qi Y, Peng S et al (1995) Epithelial cells display separate receptors for papillomavirus VLPs and for soluble L1 capsid protein. Virology 216:35–45

Ratner A, Hippe K et al (2006) Epithelial cells are sensitive detectors of bacterial pore-forming toxins. J Biol Chem 281:12994–12998

Reizes O, Benoit S et al (2008) The role of syndecans in the regulation of body weight and synaptic plasticity. Int J Biochem Cell Biol 40:28–45

Reyes-del Valle J, Chávez-Salinas S et al (2005) Heat shock protein 90 and heat shock protein 70 are components of dengue virus receptor complex in human cells. J Virol 79(8):4557–4567

Richards R, Lowy D et al (2006) Cleavage of the papillomavirus minor capsid protein, L2, at a furin consensus site is necessary for infection. Proc Natl Acad Sci USA 103(5):1522–1527

Robson K, Frevert U et al (1995) Thrombospondin-related adhesive protein (TRAP) of Plasmodium falciparum: expression during sporozoite ontogeny and binding to human hepatocytes. EMBO J 14:3883–3894

Roden R, Kirnbauer R et al (1994a) Interaction of papillomaviruses with the cell surface. J Virol 68:7260–7266

Roden R, Weissinger E et al (1994b) Neutralization of bovine papillomavirus by antibodies to L1 and L2 capsid proteins. J Virol 68:7570–7574

Roden R, Hubbert N et al (1995) Papillomavirus L1 capsids agglutinate mouse erythrocytes through a proteinaceous receptor. J Virol 69:5147–5151

Roden R, Day P et al (2001) Positively charged termini of the L2 minor capsid protein are necessary for papillomavirus infection. J Virol 75:10493–10497

Rose R, Bonnez W et al (1993) Expression of human papillomavirus type 11 L1 protein in insect cells: in vivo and in vitro assembly of viruslike particles. J Virol 67:1936–1944

Rose F, Dahlem G et al (2002) Mediator generation and signaling events in alveolar epithelial cells attacked by *S. aureus* alpha-toxin. Am J Physiol Lung Cell Mol Physiol 282:L207–L214

Rostand K, Esko J (1997) Microbial adherence to and invasion through proteoglycans. Infect Immun 65:1–8

Rusnati M, Presta M (2002) HIV-1 Tat protein: a target for the development of anti-AIDS therapies. Drug Fut 27:481–493

Rusnati M, Coltrini D et al (1997) Interaction of HIV-1 Tat protein with heparin. Role of the backbone structure, sulfation, and size. J Biol Chem 272:11313–11320

Rusnati M, Tulipano G et al (1999) Multiple interactions of HIV-I Tat protein with size-defined heparin oligosaccharides. J Biol Chem 274:28198–28205

Ryman K, Gardner C et al (2007) Heparan sulfate binding can contribute to the neurovirulence of neuroadapted and nonneuroadapted Sindbis viruses. J Virol 81:3563–3573

Salanti A, Staalsoe T et al (2003) Selective upregulation of a single distinctly structured var gene in chondroitin sulphate A-adhering Plasmodium falciparum involved in pregnancy-associated malaria. Mol Microbiol 49:179–191

Salanti A, Dahlback M et al (2004) Evidence for the involvement of VAR2CSA in pregnancy-associated malaria. J Exp Med 200:1197–1203

Schulze A, Gripon P et al (2007) Hepatitis B virus infection initiates with a large surface protein-dependent binding to heparan sulfate proteglycans. Hepatology 46:1759–1768

Selinka H, Giroglou T et al (2003) Further evidence that papillomavirus capsids exist in two distinct conformations. J Virol 77:12961–12967

Serruto D, Spadafina T et al (2010) Neisseria meningitidis GNA2132, a heparin-binding protein that induces protective immunity in humans. Proc Natl Acad Sci USA 107(8):3770–3775

Shafti-Keramat S, Handisurya A et al (2003) Different heparan sulfate proteoglycans serve as cellular receptors for human papillomaviruses. J Virol 77:13125–13135

Sharma S, Pathak S (2008) Malaria vaccine: a current perspective. J Vector Borne Dis 45:1–20

Sheehan B, Kocks C et al (1994) Molecular and genetic determinants of the Listeria monocytogenes infectious process. Curr Top Microbiol Immunol 192:187–216

Southern T, Jolly C et al (2007) EnP1, a microsporidian spore wall protein that enables spores to adhere to and infect host cells in vitro. Eukaryot Cell 6:1354–1362

Spear P (2004) Herpes simplex virus: receptors and ligands for cell entry. Cell Microbiol 6:401–410

Spillmann D (2001) Heparan sulfate: anchor for viral intruders? Biochimie 83(8):811–817

Stringer S (2006) The role of heparan sulphate proteoglycans in angiogenesis. Biochem Soc Trans 34:451–453

Suarez M, Gonzalez-Zorn B et al (2001) A role for ActA in epithelial cell invasion by Listeria monocytogenes. Cell Microbiol 3:853–864

Subramanian S, Fitzgerald M et al (1997) Regulated shedding of syndecan-1 and -4 ectodomains by thrombin and growth factor receptor activation. J Biol Chem 272:14713–14720

Talarico L, Damonte E (2007) Interference in dengue virus adsorption and uncoating by carrageenans. Virology 363(2):473–485

Taylor D, Whitehouse I et al (2009) Glypican-1 mediates both prion protein lipid raft association and disease isoform formation. PLoS Pathog 5(11):e1000666

Tkachenko E, Lutgens E et al (2004) Fibroblast growth factor 2 endocytosis in endothelial cells proceed via syndecan-4-dependent activation of Rac1 and a Cdc42-dependent macropinocytic pathway. J Cell Sci 117:3189–3199

Tristan A, Benito Y et al (2009) The signal peptide of Staphylococcus aureus panton valentine leukocidin LukS component mediates increased adhesion to heparan sulfates. PLoS One 4(4):e5042

Trybala E, Bergstrom T et al (1998) Interaction between pseudorabiesvirus and heparin/heparan sulfate. Pseudorabies virus mutants differ in their interaction with heparin/heparan sulfate when altered for specific glycoprotein C heparin-binding domain. J Biol Chem 273:5047–5052

Tuve S, Wang H et al (2008) Role of Cellular Heparan Sulfate Proteoglycans in Infection of Human Adenovirus Serotype 3 and 35. PLoS Pathog 4:e1000189

Tyagi M, Rusnati M et al (2001) Internalization of HIV-1 Tat requires cell surface heparan sulfate proteoglycans. J Biol Chem 276(5):3254–3261

Unckell F, Streeck R et al (1997) Generation and neutralization of pseudovirions of human papillomavirus type 33. J Virol 71:2934–2939

Urbinati C, Bugatti A et al (2005) a(v)β3-integrin-dependent activation of focal adhesion kinase mediates NF-B activation and motogenic activity by HIV-1 Tat in endothelial cells. J Cell Sci 118:3949–3958

Urbinati C, Nicoli S et al (2009) HIV-1 Tat and heparan sulfate proteoglycan interaction: a novel mechanism of lymphocyte adhesion and migration across the endothelium. Blood 115(15): 3335–3342

Utt M, Danielsson B et al (2001) *Helicobacter pylori* vacuolating cytotoxin binding to a putative cell surface receptor, heparan sulfate, studied by surface plasmon resonance. FEMS Immunol Med Microbiol 30:109–113

Veeltil M, Sharma-Walia N et al (2006) RhoA-GTPase facilitates entry of Kaposi's sarcoma-associated herpesvirus into adherent target cells in a Src-dependent manner. J Virol 80: 11432–11446

Veiga E, Cossart P (2005) Listeria hijacks the clathrin-dependent endocytic machinery to invade mammalian cells. Nat Cell Biol 7:894–900

Vijayvargia R, Suresh C et al (2004) Functional form of Caveolin-1 is necessary for the assembly of alpha-hemolysin. Biochem Biophys Res Commun 324:1130–1136

Vlasak M, Goesler I et al (2005) Human rhinovirus type 89 variants use heparan sulfate proteoglycan for cell attachment. J Virol 79:5963–5980

Vogt AM, Pettersson F et al (2006) Release of sequestered Malaria parasites upon injection of a glycosaminoglycan. PLoS Pathog 2(9):e100

Vollmer P, Walev I et al (1996) Novel pathogenic mechanism of microbial metalloproteinases: liberation of membrane-anchored molecules in biologically active form exemplified by studies with the human interleukin-6 receptor. Infect Immun 64:3646–3651

Volpers C, Schirmacher P et al (1994) Assembly of the major and the minor capsid protein of human papillomavirus type 33 into virus-like particles and tubular structures in insect cells. Virology 200:504–512

Volpers C, Unckell F et al (1995) Binding and internalization of human papillomavirus type 33 virus-like particles by eukaryotic cells. J Virol 69:3258–3264

Walev I, Vollmer P et al (1996) Pore-forming toxins trigger shedding of receptors for interleukin 6 and lipopolysaccharide. Proc Natl Acad Sci USA 93:7882–7887

Wang J, Tian C et al (2008) Increased levels of soluble syndecan-1 in the subretinal fluid and the vitreous of eyes with rhegmatogenous retinal detachment. Curr Eye Res 33:101–107

Weiland M, Palm J et al (2003) Characterisation of alpha-1 giardin: an immunodominant *Giardia lamblia* annexin with glycosaminoglycan-binding activity. Int J Parasitol 33:1341–1351

Westendorp M, Frank R et al (1995) Sensitization of T cells to CD95-mediated apoptosis by HIV-1 Tat and gp120. Nature 375:497–500

Woods D, Cryz S et al (1982) Contribution of toxin A and elastase to virulence of *Pseudomonas aeruginosa* in chronic lung infections of rats. Infect Immun 36:1223–1228

Xiao H, Neuveut C et al (2000) Selective CXCR4 antagonism by Tat: implications for in vivo expansion of coreceptor use by HIV-1. Proc Natl Acad Sci USA 97:11466–11471

Yang R, Day P et al (2003) Cell surface-binding motifs of L2 that facilitate papillomavirus infection. J Virol 77:3531–3541

Yeager M, Aste-Amezaga M et al (2000) Neutralization of human papillomavirus (HPV) pseudovirions: a novel and efficient approach to detect and characterize HPV neutralizing antibodies. Virology 278:570–577

Zautner A, Komer U et al (2003) Heparan sulfates and coxsackievirus-adenovirus receptor; each one mediates coxcackievirus B3 PD infection. J Virol 77:10071–10077

Zautner A, Jahn B et al (2006) N- and 6-O-sulfated heparan sulfates mediate internalization of Coxsackievirus B3 variant PD into CHO-K1 cells. J Virol 80:6629–6636

Zeng Y, Zhang X et al (2007) Intracellular Tat of human immunodeficiency virus type 1 activates lytic cycle replication of Kaposi's sarcoma-associated herpesvirus: role of JAK/STAT signaling. J Virol 81:2401–2417

Zhang J, Stephens R (1992) Mechanism of *C. trachomatis* attachment to eukaryotic host cells. Cell 69:861–869

Zvibel I, Halfon P et al (2009) Syndecan 1 (CD138) serum levels: a novel biomarker in predicting liver fibrosis stage in patients with hepatitis C. Liver Int 29:208–212

Chapter 3
Glycans in Cancer

Lubor Borsig

Abstract Alteration in cellular glycosylation is a common phenotypic change associated with malignant transformation and cancer progression. Cell surface oligosaccharides carried on glycoproteins and glycolipids mediate communication among cells and facilitate cell adhesion, processes that are central during cancer progression. Accumulating evidence indicates that glycans contribute to tumor invasion, metastasis, and angiogenesis. A better understanding of glycan function during malignancy can help designing strategies for antitumor therapies, either alone or in combination with other standard treatments. This chapter gives an overview on cancer-specific changes of glycosylation on O- and N-glycans, with the focus on the function of these oligosaccharides in cancer progression.

3.1 Introduction

Changes in glycosylation are associated with several physiological and pathophysiological processes, including cell growth, migration, differentiation, host–pathogen interactions, and those occurring during tumorigenesis. Several glycans on both the tumor and host surfaces have been identified as mediators of various processes during tumor progression (Fuster and Esko 2005; Kannagi 1997; Kim and Varki 1997; Kobata and Amano 2005). Tumor progression is characterized by several rather specific alterations in cellular signaling. These can either result in changes of cell cycle and an enhanced cell proliferation or lead to reduction of cell–cell interactions within the tumor environment and thereby promoting tumor cell invasiveness. While enhanced invasiveness of metastatic cells requires efficient tumor cell dissociation from the primary tumors, partly the same altered glycan structures can mediate interactions of disseminating tumor cells with leukocytes, platelets, and endothelial cells of lymphatic or blood vasculature, thereby facilitating the metastatic spread.

L. Borsig
Institute of Physiology, University of Zürich, Zürich 8057, Switzerland
e-mail: lborsig@access.uzh.ch

M.S.G. Pavão (ed.), *Glycans in Diseases and Therapeutics*, Biology of Extracellular Matrix, 63
DOI 10.1007/978-3-642-16833-8_3, © Springer-Verlag Berlin Heidelberg 2011

The role of glycans in cancer progression was suggested early on when many monoclonal antibodies raised against various carcinomas often recognized short oligosaccharide structures usually not found in the healthy tissues (Dennis and Laferte 1987; Feizi 1985; Hakomori 1986). A correlation between altered glycosylation and poor prognosis of cancer patients with various carcinomas further raised interest in tumor-specific glycosylation mainly for its potential diagnostic use (Kannagi 1997; Kim and Varki 1997). Indeed, many tumor markers currently used in clinics for early diagnosis or monitoring of cancer progression are glycoproteins detected in the serum, for example, CEA, CA19-9, CA125, etc. (Hollingsworth and Swanson 2004; Rhodes 1999).

3.2 Glycosylation During Cancer

3.2.1 Synthesis of Glycans

Glycans encompasses oligosaccharides that can exist either in a free form or in large complex carbohydrate molecules attached to proteins or lipids. Based on the bond, glycoconjugates can be divided into glycoproteins, glycolipids, and glycosaminoglycans. Glycans are mostly found on a cell surface, in a extracellular matrix, and in various cellular compartments such as: endoplasmic reticulum (ER); the Golgi apparatus; lysosomes; and nuclei. In fact, glycosylation is one of the most common posttranslational modifications of proteins, while almost half of them are glycosylated. The synthesis and the attachment of glycans to proteins or lipids occur in the secretory pathway (ER, Golgi apparatus). Glycan sequence is not directly coded by a DNA template. The structure specificity of glycans relies on the expression of various proteins such as glycosyltransferases, glycosidases, enzymes involved in synthesis of activated sugars and sugar transporters. Finally, the spatial presence of the specific enzymes in individual compartments of the Golgi apparatus is required for an efficient glycans synthesis. On a contrary to DNA and protein synthesis which is composed of linear polymers, the structure of glycans is defined not only on the position and the type of the linkage (alpha or beta linkage) between two carbohydrate units but also by the fact that two or more sugar units can be linked to the underlying monosaccharide unit (glycan branching). Based on the initial attachment of carbohydrates to proteins, we recognize two major types of protein glycosylation (a) N-linked glycosylation, where the glycan is covalently bound to the Asn residue and (b) O-linked glycosylation, where the glycan is attached to either Ser or Thr residue. While potential glycosylation sites for N-glycans has been identified (Asn–X–Ser/Thr, while X cannot be Pro), there is no single consensus sequence for O-linked glycan attachment, and the regulation of O-glycans is rather compound. Mucins are proteins with a large number of O-linked glycans, which may comprise 80% of the molecule by weight (Varki et al. 2008). Besides glycoproteins, there are different glycosaminoglycans (e.g., heparan sulfate) that exist either as linear polysaccharide polymers or as parts of proteoglycans (discussed in Chap. 6) and glycolipids that consist of oligosaccharides linked to sphingolipids – ceramides.

3.2.2 Alteration of Glycosylation During Cancer

Under normal physiological condition, a single glycoprotein is synthesized in a range of glycoforms that depends on the occupancy of potential glycosylation sites and the structure of glycans. Depending on the location, glycans are involved in the modulation of cell adhesion, motility, and various signaling events (Fuster and Esko 2005). In pathophysiological situations, during malignancy in particular, numerous alterations in glycan structures have been documented (e.g., Kannagi 1997; Kobata and Amano 2005). Although the functional consequence of these alterations is still under investigation, there is accumulating evidence that changes in glycosylation allow malignant cells to control cell adhesion, cell motility, and affects cell growth and proliferation, all of which are associated with cancer progression. There is a general agreement that tumor cell arise from genetic alterations in proto-oncogenes and tumor suppressors. Oncogene activation and loss of tumor cell suppressors result in the observed changes of glycosylation with functional consequences mostly during cancer progression, particularly by cell invasion and metastasis. However, cancer glycans were also shown to affect cell proliferation and tumor growth (Girnita et al. 2000).

Despite numerous possibilities how glycans can be formed there is a rather limited number of structures that correlate with cancer progression and poor prognosis (Table 3.1). The first group of cancer-associated glycans encompasses

Table 3.1 Common glycan alterations associated with cancer progression of carcinoma

Glycan modification	Change in biosynthesis	Proposed major function	Clinical correlation	References
Increased β6-branching on N-glycans	Increased GnT-V activity	Altered adhesion to ECM enhanced migration	Breast, colon	Granovsky et al. (2000)
Increased sialylation	Increased sialyltransferase activity	Decreased cell–cell interactions, enhanced motility and invasiveness	Lung, breast	Pinho et al. (2007), Ogata et al. (1998)
Increased expression of $sLe^{x/a}$ structures	Altered expression of fucosyltransferases, sialyltransferases	Enhanced tumor cell adhesion during metastasis	Colon, lung	Khatib et al. (1999), Läubli et al. (2006)
Increased expression of Tn and sialyl-Tn structures	Decreased core 2 activity altered sialyltransferase and pp GalNAc activity	Enhanced motility	Colon, breast	Shibao et al. (2002), Ogata et al. (1998), Pinho et al. (2007)

truncated shorter structures of normally occurring glycans that are not found in healthy tissues, for example, Tn, sialyl-Tn, and T antigens (Brockhausen et al. 1998; Taylor-Papadimitriou et al. 1999). Reduced or missing β1,3-galactosyltrasferase activity that enables the extension of the mucin type core 1 glycan was identified as the main reason for the enhanced expression of Tn and sialyl-Tn structures by various carcinomas (Brockhausen et al. 1998). Reduced O-glycosylation resulting in short glycans leads to exposure of peptide sequences that are not present in healthy tissues (Brockhausen 2006; Kim et al. 1996; Taylor-Papadimitriou et al. 1999). Accordingly, a specific antitumor targeting with antibody to cancer specific-mucin epitopes, MUC1, has been already therapeutically tested (Tang et al. 2008). The second major group of cancer-associated glycans is represented by enhanced expression of unique glycans structures that are predominantly terminal oligosaccharide structures, such as sialyl Lewis$^{x/a}$, changes in terminal sialylation and fucosylation, and glycan branching (Kannagi 1997).

3.2.3 Alterations of N-Linked Glycosylation in Cancer

3.2.3.1 N-Linked Glycosylation

Many glycoproteins are modified with N-linked glycans that were shown to affect protein folding, cell–cell interactions, and cell adhesion (Varki et al. 2008). Most receptors on the cell surface are N-glycosylated, including integrins, epithelial growth factor receptor (EGFR), and transforming growth factor β receptor (TGFβR). Similarly, several secreted molecules such as hormones and growth factors bind to cell surfaces through carbohydrate receptors lectins, for example, Galectins and Siglecs (Crocker et al. 2007; Lau and Dennis 2008). Therefore, changes of N-glycan glycosylation on cancer cells may affect several processes during cancer progression such as tumor growth, tumor cell adhesion, and tumor invasion (Table 3.1).

3.2.3.2 Biosynthesis of N-Glycans

The biosynthesis of N-glycans can be divided into four phases based on different compartments of the cellular secretory pathway (Kornfeld and Kornfeld 1985; Varki et al. 2008) (1) transfer of the preassembled core oligosaccharide structure from a dolichol-linked donor to the protein that occurs in the endoplasmic reticulum; (2) reduction/trimming of the transferred core glycan by the action of glycosidases in the ER and proximal Golgi apparatus; (3) branching/addition of *N*-acetylglucosamine (GlcNAc) residues to the processed structure in the Golgi apparatus; and (4) completion of the glycan structure by addition of galactose, fucose, and sialic acid in the distal Golgi apparatus and the trans-Golgi network. Cancer-specific N-linked glycan structures are largely a result of changes in branching and completion of the glycan biosynthesis.

3.2.3.3 N-Glycan Branching: The Role of GnT-V

The addition of GlcNAc is an initial step in the synthesis of a complex N-linked glycans. There are six different β-*N*-acetylglucosaminyltransferases (GnTs) which are responsible for the synthesis of N-linked glycans in mammals (Fig. 3.1). Malignant transformation is accompanied by increased β1,6GlcNAc-branching on N-glycans and was related to tumor metastasis (Dennis et al. 1987; Granovsky et al. 2000). Enhanced expression of GnT-V was detected in mammary, colon, lung carcinomas, and neuroblastomas (Kobata and Amano 2005; Lau and Dennis 2008). Ectopic expression of gene encoding GnT-V in various epithelial cells resulted in increased cell motility, tumor formation, and enhanced metastasis (Chen et al. 1998; Demetriou et al. 1995). The function of β1,6GlcNAc structures during cancer progression was further assessed in GnT-V-deficient mice (Dennis et al. 2002; Granovsky et al. 2000). Mice lacking β1,6GlcNAc-modified glycans appeared normal but showed different responses to extrinsic conditions such as autoimmune disease and cancer progression (Granovsky et al. 2000). Tumor growth and metastasis were considerably decreased in the absence of GnT-V in a middle T oncogene-induced mouse mammary tumor model. The molecular mechanism of the β1,6GlcNAc contribution to metastasis was elucidated recently. Previous findings have shown that N-linked glycans on α5β1 integrins are essential for their proper binding to fibronectin (Veiga et al. 1995; Zheng et al. 1994). Increased expression of GnT-V resulted in a reduced cell adhesion to fibronectin where the more motile and less adhesive phenotype was identified to be the result of reduced cell-surface clustering of α5β1 integrins (Guo et al. 2002). Specifically, increased β1,6-branching of N-linked glycans on the β1 subunit of an integrin causes altered clustering of integrins and thereby reduced binding to fibronectin. The functional consequence of enhanced β1,6-branching on metastatic behavior of tumor cells was further

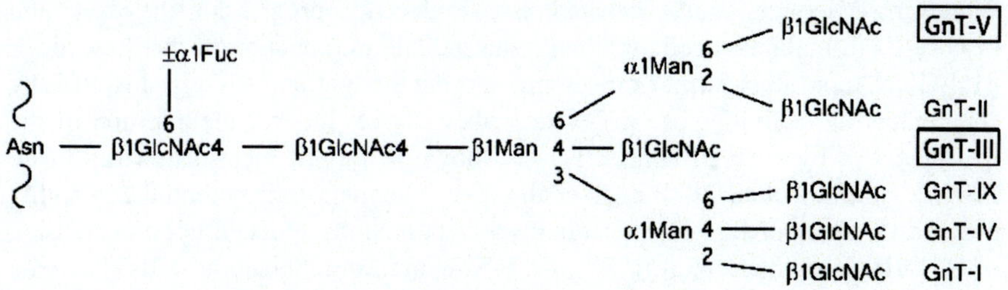

Fig. 3.1 Synthesis of N-glycans in the Golgi apparatus. There are six β-*N*-acetylglucosaminyl-transferases (GnT) responsible for the branching of a complex type of N-glycans. GnT-V-mediated β1,6-branching has been linked to metastasis. GnT-III β1,4-branching activity is detected in epithelial, nontumorigenic tissues and has been shown to suppress cancer progression by counter-acting of GnT-V activity

linked to alterations of growth factor receptors on tumor cell surfaces (Partridge et al. 2004). The analysis of mammary epithelial tumor cells from GnT-V-deficient mice have shown reduced responsiveness to several cytokines, including epithermal growth factor (EGF), insulin-like growth factor (IGFR), and platelet-derived growth factor (PDGF) when compared to GnT-V expressing tumor cells. GnT-V expression was found to be the rate-limiting step for cytokine signaling and subsequently for cell motility and tumor metastasis. Limited N-glycan branching of epidermal growth factor and transforming growth factor-β receptors has been observed in the absence of GnT-V. As a result, enhanced cell surface clearance of receptors by endocytosis caused a reduced responsiveness to cytokines (Partridge et al. 2004). Galectin-3, an endogenous lectin, which preferably binds to poly-*N*-acetyllactosamine carried mostly on β1,6-branched N-glycans was identified as a key mediator of a prolonged cell surface retention of growth factor receptors (Partridge et al. 2004). The interaction of glycoproteins carrying N-glycans, for example, EGFR, IGFR, PDGFR, with endogenous lectins is critical for the cellular adaptation during cancer (Dennis et al. 2009). Furthermore, branching of N-glycans that occurs in the Golgi apparatus has been shown to be sensitive to the concentration of activated GlcNAc, which serves as a donor for GnT (Lau et al. 2007).

Downregulation of GnT-V expression by siRNA in a highly metastatic and invasive human breast carcinoma cell line, MD-MB321, resulted in a suppressed EGF-downstream signaling and invasiveness-related phenotype (Guo et al. 2002). Knock-down of GnT-V did not alter the expression levels of EGFR, but lowered expression of N-linked β1,6-branching on this receptor. Similarly, altered N-glycosylation of EGFR caused by overexpression of GnT-V resulted in an increased EGF-mediated MAPK signaling (Guo et al. 2004). A general inhibition of N-glycosylation was shown to downregulate cell surface expression of IGFR in Ewing's sarcomas and to limit tumor cell growth and survival (Girnita et al. 2000). These findings indicated that inhibition of N-glycan synthesis could be explored for its therapeutic potential.

3.2.3.4 GnT-III Versus GnT-V

While GnT-V expression has been associated with cancer progression, the expression of GnT-III in malignant cells is often reduced. The introduction of the "bisecting" β1,4-GlcNAc inhibits further processing and the elongation of N-glycans, thereby counteracts the activities of GnT-V described above. Indeed, the addition of the β1,4-GlcNAc by GnT-III (Fig. 3.1) was shown to inhibit metastasis (Isaji et al. 2004; Yoshimura et al. 1995; Zhao et al. 2006). The metastatic potential of a highly metastastic clone of B16-F1 melanona was significantly reduced upon expression of GnT-III (Yoshimura et al. 1995). GnT-III-expressing melanoma cells showed a reduced cell spreading, migration on fibronectin, and alteration of α5β1 integrin glycosylation that resulted in downregulation of integrin-mediated signaling (Isaji et al. 2004). A direct evidence for the competition between GnT-III and GnT-V and its effect on cancer progression was recently shown in human gastric cancer cell line MKN45 (Zhao et al. 2006). GnT-III-induced modification of α3β1 integrin inhibited the GnT-V-stimulated cell migration on laminin 5, and this reduced migration was directly linked to the increased presence of a bisecting GlcNAc on the α3 integrin subunit (Zhao et al. 2006). Interestingly, increased expression of GnT-III has been detected in dense epithelial cell layers and correlated with the expression of E-cadherin, an adhesion molecule associated with reduced invasiveness and reduced metastasis of tumor cells (Iijima et al. 2006). However, the contribution of GnT-III to maintenance of epithelial phenotype and to modulation of E-cadherin expression remains to be elucidated.

An additional aspect of N-glycan synthesis that is associated with cancer progression is the enhanced expression of terminal glycan structures such as a tetrasaccharide NeuAcα2,3Galβ1,4(Fucα1,3)GlcNAc, also called sialyl Lewis^x (sLe^x). It has been reported that sLe^x structures are mostly found on extended β1,6GlcNAc N-branches carrying *N*-acetylactosamine (Galβ1,3/4GlcNAc) (Fukuda et al. 1984). The function of sLe $^{x/a}$ structures during metastasis will be discussed later.

3.2.4 Alterations of O-Linked Glycosylation in Cancer

3.2.4.1 O-Linked Glycosylation and Carcinoma Mucins

O-linked glycans are found on many glycoproteins including mucins and several membrane receptors. Mucins are cell-surface bound or secreted heavily O-glycosylated glycoproteins produced by various epithelial cell types. The structurally common feature of all mucins is a tandem-repeat domain that contains similar sequences rich in serine, threonine, and proline residues (Brockhausen 1999; Hollingsworth and Swanson 2004). O-linked oligosaccharides are built on threonine and serine residues in these domains. The number and the exact sequence of these domains vary between different mucins. There are about 20 genes encoding mucins with broad but epithelium-specific expression that fulfill different physiological functions. However, several mucins were associated with cancer progression including cell-surface mucins MUC1, MUC4 and secreted gel-forming mucin MUC2 (Byrd and Bresalier 2004; Hollingsworth and Swanson 2004; Taylor-Papadimitriou et al. 1999).

Cancer cells, carcinomas in particular, express altered glycans on mucins. Cancer-specific O-glycans arise from a deregulation of mucin gene expression and the enzymes involved in O-glycan synthesis. The acquired novel glycan structures on carcinoma mucins enable cancer cells to bind to a variety of different receptors and thereby modulate the biological properties of tumor cells.

3.2.4.2 Biosynthesis of O-Glycans

The initial step of O-glycosylation is the addition of *N*-acetylgalactosamine (GalNAc) from an activated sugar donor (UDP-GalNAc) to the hydroxyl group on Thr or Ser residue. Polypeptide GalNAc-transferases (pp GalNAc transferases), a family of about 20 different enzymes, controls the initiation of O-glycosylation (Ten Hagen et al. 2003). These enzymes show overlapping but different specificities with regard to cell types and a defined peptide sequence. The specificity and the functional requirement for pp GalNAc-transferases are still under investigation and most of the available data about O-glycosylation site-specificity are derived from in vitro studies (Tarp and Clausen 2008). Furthermore, O-glycan structures already attached to a mucin can affect the utilization of adjacent O-glycosylation sites making the definition of the peptide motif recognized for the O-glycan attachment rather complex. Despite the critical role of pp GalNAc-transferases by initiation of O-glycosylation, there is limited knowledge about their involvement in cancer progression and cancer-associated changes in glycosylation. Nevertheless, there is evidence that expression of pp GalNAc-transferase 6 correlates with metastasis of breast and gastric carcinomas (Freire et al. 2006; Gomes et al. 2009). Histological evaluation of colon carcinomas revealed a positive correlation between pp GalNAc-transferase 3 expression and a good prognosis for patients (Shibao et al. 2002). Also high levels of GalNAc-T12 are found in the normal colon mucosa. Inactivation of pp GalNAc-transferase 12 (GalNAc-T12), caused by a germ-line mutation, was identified in colon cancer patients (Guda et al. 2009). This observation suggests that altered glycosylation caused by certain genetic defects in the O-glycosylation pathway may contribute to cancer progression.

3.2.4.3 Carcinoma-Associated Alterations of O-Glycans

Under normal physiological condition, mucin glycans are modified by an addition of a galactose on the initial GalNAc residue to form a core 1 structure (Fig. 3.2). This disaccharide structure, Galβ1,3–GalNAc, acts as a substrate for the β1,6GlcNAc-transferase that forms the core 2 structure (Fig. 3.2). Typically, normal mucins of a mammary gland carries polylactosamine disaccharides (Galβ1,4–GlcNAc) on the β1,6GlcNAc branch, which are further modified with a fucose (Hanisch 2001). In a normal colonic mucosa, polylactosamine chains are built on the core 3 structure that is synthesized by a β1,3GlcNAc-transferase on a core 1 structure (Fig. 3.2).

During malignant transformation, a profound change in the structure of O-glycans was found in all examined carcinomas. Truncation of O-glycans and the presentation of sialyl Lewis antigens are the most widely observed alterations in carcinomas (Brockhausen 2006). MUC1 is the major mucin detected in carcinomas of the breast or colon. MUC1 mucin in breast carcinomas carries mostly truncated O-glycans such as Tn and sialyl-Tn-structures (Fig. 3.2). An increased presence of these glycans has been linked to the absence of core 2 GlcNAc-transferase activity that results in an accumulation of core 1-based structures that are further converted to sialyl Tn or sialyl-T antigens by sialyltransferases (Brockhausen 2006). Sialylation of the core 1 structure is often found on mammalian carcinomas that is due to an increased expression of sialyltransferases (STs) adding sialic acid either in α2,3 or α2,6 linkage (Hanisch 2001). A major but not a complete shift from core 2-based

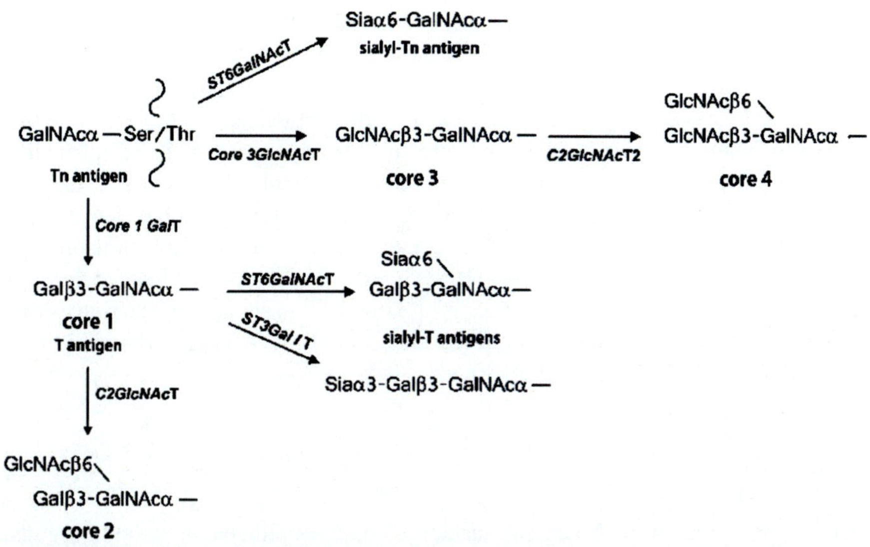

Fig. 3.2 Biosynthesis of O-glycans. O-glycosylation is initiated by an addition of a GalNAc residue to the protein. This simplest O-glycan structure is called Tn-antigen. Tn-antigen can be further converted to core 1 by β1,3-Galactosyltransferase (Core 1 GalT) or to core 3 by β1, 3-GlcNAc-transferase (Core 3GlcNAcT). Core 1 can be branched by C2GlcNAcT to form core 2, which is commonly found in various epithelia, including mammary tissues. Core 4 based glycans are common in colonic mucosa. During malignant transformation truncated O-glycans such as Tn and T structures are commonly observed on carcinoma mucins. These glycans are further modified by sialyltransferases that result in sialyl Tn and sialyl T antigens

O-glycans toward short core 1 glycans is typical for breast carcinomas. Despite an overall decrease of core 2 structures, an increase in sLe^x antigens has been detected (Fig. 3.2) that is likely ascribed to the elevated levels of fucosyltransferase activity (Brockhausen 2006).

In colorectal cancer, a shift toward short glycans structures like Tn and sialyl-Tn epitopes has been also detected. This is a result of nearly absent levels of a core 3 β3-GlcNAc-transferase in colon carcinomas (Brockhausen 1999). Furthermore, a common effect in many tumors is the loss of a chaperone COSMC, which is required for the formation of an active form of β1,3-galactosyltransferase that acts on the GalNAc residue, and forms the core 1 structure (Ju et al. 2008). Lack of COSMC results in accumulation of Tn antigen, which is mostly converted to sialyl-Tn. Expression of α2,6Sia-Tn antigens was correlated with a poor prognosis in colorectal cancer (Shimodaira et al. 1997).

Besides short O-glycans, there is a significant increase in presence of sialyl Lewis x and sialyl Lewis a (its isomer), $sLe^{x/a}$, structures on colon carcinomas. These terminal structures are built on a core 2 structure that is synthesized by one of the two different core 2 β6-GlcNAc-transferases found to be maintained in colon carcinomas (Shimodaira et al. 1997). Elevated expression of $sLe^{x/a}$ on colon carcinomas is consistently associated with poor prognosis due to enhanced metastasis (Nakamori et al. 1993; Okazaki et al. 1998).

3.2.4.4 Carcinoma Mucins and Cancer Progression

Physiological functions of mucins in normal and malignant cells are highly complex. Mucin layers on epithelium represent a physical barrier against microorganisms and other materials. Mucins and mucus gels, composed of secreted mucins associated with the epithelial surfaces, can bind active molecules like cytokines and growth factors, and thereby contribute to recruitment and migration of leukocytes. Depending on the structures of O-glycans, mucins can facilitate antiadhesive functions or mediate cell adhesion (Brockhausen 2006). Finally, recent findings provide evidence that cell-surface mucins are involved in signal transduction events in response to the environment changes and altered glycosylation (Ahmad et al. 2007; Bitler et al. 2009; Gao et al. 2009).

3.2.4.5 MUC1

MUC1 is expressed at the apical surface of ductal epithelia of various organs, including breast, airways, pancreas, and the gastrointestinal tract. It is well-established that the cell-surface associated MUC1 mucin is overexpressed and contains altered glycosylation structures in several different carcinomas (Brockhausen et al. 1995; Hollingsworth and Swanson 2004). There is substantial evidence that MUC1 contributes to metastasis of tumors, but its effect in different types of cancer may vary (Hollingsworth and Swanson 2004). Enhanced expression of MUC1 was found in colon cancer patients with metastasis compared to patients with localized tumors (Nakamori et al. 1994). Targeted deletion of a murine MUC1 significantly reduced tumor growth of an oncogene-induced mammary tumor and mildly affected metastasis (Spicer et al. 1995). In colon and pancreatic carcinomas, MUC1 mucins are rich in $sLe^{x/a}$ and sialyl-Tn structures (Burdick et al. 1997). A clear association between an enhanced $sLe^{x/a}$ expression and poor prognosis of carcinoma patients argues for a role of these glycans in cancer progression (Jorgensen et al. 1995; Nakamori et al. 1997a; Ogawa et al. 1997; Renkonen et al. 1997; Tatsumi et al. 1998; Tozawa et al. 2005), will be discussed later in this chapter.

There is accumulating evidence that MUC1 mucin contributes to regulation of cell proliferation and signal transduction (Ahmad et al. 2007; Bitler et al. 2009; Gao et al. 2009). Carcinoma-associated overexpression of MUC1 has been associated with constitutive activation of p65 subunit of NF-κB (Ahmad et al. 2007). A direct interaction between MUC1 and IkappaB kinase complex indicated that MUC1 confers sustained activation of the NF-κB pathway in human cancers. The cytoplasmic tail of MUC1 binds to β-catenin and associates with glycogen synthase kinase 3β (GSK3β) and the phosphorylation of MUC1 resulted in a decreased β-catenin binding (Li et al. 1998). As a result, the overexpression of MUC1 in the absence of GSK3β activity inhibits formation of the E-cadherin–β-catenin complex, one of the known mechanisms contributing to cell motility and metastasis (Onder et al. 2008). Furthermore, the cytoplasmic tail of MUC1 mucin inhibits the degradation of the epidermal growth factor receptor (EGFR), thereby enhances tumor cell transformation and metastasis (Bitler et al. 2009). Knockdown of MUC1 in lung cancer cells resulted in inhibition of a tumor growth and metastasis that was associated with enhanced apoptosis and inhibition of cell proliferation (Gao et al. 2009). MUC1 gene was identified as one of downstream targets of STAT3, which is aberrantly activated in human cancers including lung cancer. It has also been shown that tandem-repeat domains in the extracellular parts of MUC1 transduce extracellular cues to intracellular signals through the cytoplasmic tail (Kohlgraf et al. 2003). Taken together, these findings indicate that MUC1 may serve as a sensor of the local environment and has the potential to provide signals that contribute to metastatic characteristic of tumor cells.

3.2.4.6 MUC2

MUC2 is a major secreted gastrointestinal mucin. Several studies have reported that MUC2 expression is decreased in colon carcinomas (Baldus et al. 2002; Chang et al. 1994). Targeted inactivation of MUC2 gene resulted in frequent spontaneous development of adenocarcinomas in the small intestine, colon, and rectum of mice (Velcich et al. 2002). The lack of MUC2 mucin leads to the absence of a movable mucus layer in the gastrointestinal tract, thereby exposing epithelium to a direct contact with bacteria (Johansson et al. 2008). Although the exact role of MUC2 in tumor suppression remains to be defined, the protective role of MUC2-composed mucus in the gastrointestinal tract further supports a link between chronic inflammation and cancer development (Coussens and Werb 2002; Karin and Greten 2005; Mantovani et al. 2008).

3.2.4.7 MUC4

Enhanced expression and altered glycosylation of MUC4 mucin has been reported in various cancers, including pancreatic and lung cancers. Since MUC4 expression was found to be downregulated in prostate and salivary gland cancers, MUC4 mucin may have diverse functions that are dependent on a local environment and glycan alterations of MUC4 (Chaturvedi et al. 2008). One of the hallmarks of epithelial transformation is the loss of cell polarity. Thus, mucins, normally present or secreted on the apical side of epithelial cells, can gain novel functions. Accordingly, MUC4 in pancreatic cancer has been shown to interact with tyrosine receptor kinase ErbB2, and thereby potentiates signaling through an epidermal growth factor-like domain in its extracellular region (Senapati et al. 2008). The interaction of MUC4 with ErbB2 was found to be independent of O-glycan domains (Workman et al. 2009). Finally, MUC4 expression in melanoma and breast cancer cells contributes to inhibition of apoptosis and resistance to chemotherapy (Workman et al. 2009).

3.2.5 *Altered Glycosylation Affects Biology of Cancer*

3.2.5.1 Altered Glycosylation and Metastasis

During metastasis, malignant cells detach from the primary tumor and migrate through the surrounding tissue, enter the blood circulation or lymphatic system, extravasate and establish new metastatic colonies. In different tissues, altered glycan structures on carcinoma mucins are capable of mediating cell adhesion, cell detachment, and communication within a tumor environment (Fuster and Esko 2005; Hollingsworth and Swanson 2004). The observed loss of epithelial cell polarity during early stages of malignant transformation leads to the ubiquitous cell-surface expression of carcinoma mucins. The altered expression of carcinoma mucins can block cell–cell interactions by steric hindrance through glycans (Brockhausen 2006; Fuster and Esko 2005; Hollingsworth and Swanson 2004). The acquired capacity of tumor cells to migrate through extracellular matrix and to dissociate from the primary tumor is promoted by the expression of sialylated glycans that promotes tumor cell detachment. In particular, elevated expression of sialyl-Tn reduced cell–cell interactions and increased the metastatic potential (Julien et al. 2006; Leivonen et al. 2001). Consequently, enhanced expression of sialyl-Tn has been associated with poor prognosis in colorectal and gastric carcinomas (Ogata et al. 1998; Pinho et al. 2007).

Hematogenous metastasis is the most common route of cancer spread for carcinomas. During the intravascular phase, tumor cells interact with platelets, leukocytes, and eventually arrest in small vessels of distant organs. There is evidence that tumor cell emboli formation consisting of platelets and leukocytes potentiates tumor metastasis (Borsig et al. 2007; Nierodzik and Karpatkin 2006). Thus, tumor cell interactions within the surrounding environment (platelets, leukocytes, endothelium and stromal cells) is critical for metastasis and the altered glycosylation seems to be involved in this process (Chiang and Massague 2008; Fuster and Esko 2005; Krause and Turner 1999; Varki and Varki 2002). Altered glycosylation and the enhanced expression of sLe^x, in particular, are associated with metastasis suggesting a direct effect on this process. Histological analysis of colon carcinoma patients provided evidence that the prognosis was significantly improved in patients negative for sLe^x structures on tumor cells (Nakamori et al. 1993). Several studies of colorectal carcinomas demonstrated that sLe^x expression also correlated with the stage of the disease, disease recurrence, and the presence of lymph node metastases (Izumi et al. 1995; Nakamori et al. 1993, 1997b). The positive relationship between sLe^x expression and metastasis associated with decreased survival has been shown for various cancers, including lung (Ogawa et al. 1997), prostate (Idikio 1997; Jorgensen et al. 1995), breast (Renkonen et al. 1997; Yamaguchi et al. 1998), gastric (Tatsumi et al. 1998) and renal cell carcinomas (Tozawa et al. 2005). The accumulated clinical evidence links sLe^x expression to cancer progression, while the underlying mechanism is under investigation.

3.2.5.2 Cell–Cell Interactions During Metastasis

Tumor cell adhesion to endothelium, platelets, and leukocytes is promoted by the O-glycans on carcinomas. One family of vascular cell adhesion molecules, selectins, was shown to facilitate interactions of tumor cells with blood constituents during the hematogenous phase of metastasis (Kannagi 1997; Läubli and Borsig 2010b; Varki and Varki 2002). Physiological functions of selectins are well described in processes of inflammation, hemostasis, and wound repair (Kansas 1996). There are three members of the selectin family: P-selectin, rapidly expressed upon activation of platelets and endothelial cells, L-selectin, constitutively expressed on virtually all leukocyte populations, and E-selectin, expressed on endothelial cells upon prolonged activation. The initial steps of leukocytes adherence on activated endothelium are mediated by rapid but reversible interactions between selectins and their ligands. The main selectin ligand for all selectins is sLe^x tetrasaccharide that is normally present only on discrete number of proteins in the vasculature. The best characterized selectin ligand carrying sLe^x is P-selectin glycoprotein ligand 1 (PSGL-1) that is concentrated in the microvilli on leukocyte surface (Liu et al. 1998).

Tumor cell interactions with the endothelium of distant organs are critical for metastasis, therefore an involvement of endothelial E-selectin has been proposed (Kannagi 1997; Khatib et al. 1999). Initial evidence for E-selectin involvement in metastasis has been obtained in mice overexpressing E-selectin in a liver that diverted metastasis to this organ (Biancone et al. 1996). Since endothelial E-selectin expression requires a prolonged activation, its involvement in the early steps of metastasis may be limited. While, E-selectin inhibition resulted in attenuation of experimental liver metastasis (Khatib et al. 1999), there was no effect on experimental lung metastasis in the absence of E-selectin (Läubli and Borsig 2010a). These observations indicate that E-selectin contributes to metastasis possibly in a tissue-specific manner.

Tumor cells carrying selectin ligands present in the blood circulation may encounter vascular selectins, and thereby contribute to cancer progression. A direct evidence for selectin-mediated facilitation of metastasis has been obtained in mice deficient either in P- and/or L-selectin (Borsig et al. 2001, 2002; Brown et al. 2006; Kim et al. 1998; Läubli et al. 2006). Attenuation of metastasis of several cancer cell lines has been observed in the absence of P-selectin that resulted from a reduced/minimal platelet aggregation on tumor cells (Borsig et al. 2001; Kim et al. 1998; Läubli et al. 2006; Ludwig et al. 2004). Elimination of carcinoma mucins (carrying sLe^x structures) from tumor cells prior to the intravenous injection also reduced platelet aggregation and metastasis (Borsig et al. 2001). Attenuation of metastasis was observed also in L-selectin deficient mice, implicating leukocytes as active modulators of this process (Borsig et al. 2002). The requirement for L-selectin-dependent recruitment of neutrophils and monocytes to metastasizing tumor cells indicates that these leukocytes contribute to initial phase of metastasis and possibly enable tumor cell extravasation (Läubli et al. 2006, 2009). The absence of P-and L-selectin-mediated interactions or their specific inhibition resulted in a reduced colonization of tissues by tumor cells and was accompanied by a reduced tumor cell survival. Pharmacological inhibition of P- and L-selectin by heparin derivatives devoid of anticoagulant activity attenuated metastasis of various tumor cells in a selectin-specific manner (Hostettler et al. 2007; Kragh et al. 2005). These findings suggest that P- and L-selectin contribute to metastasis of carcinoma cells carrying selectin ligands by directly affecting initiation of metastasis. In this context, the presence of sLe^x on tumor cells is associated with an enhanced metastatic behavior and poor prognosis.

3.3 Conclusions and Perspective

Profound alterations of glycosylation contribute to cancer progression in various ways as discussed in this chapter. Hence, specific targeting of a glycosylation pathway in cancer may provide a valuable tool to affect several processes in cancer progression (Fuster and Esko 2005; Hakomori 2001). Glycan-based immunomodulators and vaccines are just two different approaches currently under development in several laboratories.

The current evidence from clinical studies strongly argues for a correlation between elevated expression of certain glycan structures, for example, $sLe^{x/a}$ and sialyl-Tn, and poor prognosis due to metastasis. Experimental data provide a direct evidence for the involvement of selectins in the early steps of a metastatic cascade, where P- and L-selectin mediate interactions between tumor cells and blood constituents (platelets, leukocytes, and endothelial cells). Targeting of selectins is currently being explored only in animal models. In contrary, heparin is being tested in a number of clinical trials as an antimetastatic treatment. Although heparin was shown to consistently attenuate metastasis by inhibition of selectin in a number of animal models (Borsig 2010), its affectivity in a clinical setting remains to be determined.

Further studies of pathogenic mechanisms that contribute to cancer progression through cancer-associated glycan structures will help us to design better therapeutic strategies to combat cancer.

References

Ahmad R, Raina D, Trivedi V, Ren J, Rajabi H, Kharbanda S, Kufe D (2007) MUC1 oncoprotein activates the IkappaB kinase beta complex and constitutive NF-kappaB signalling. Nat Cell Biol 9:1419–1427

Baldus SE, Monig SP, Hanisch FG, Zirbes TK, Flucke U, Oelert S, Zilkens G, Madejczik B, Thiele J, Schneider PM, Holscher AH, Dienes HP (2002) Comparative evaluation of the prognostic value of MUC1, MUC2, sialyl-Lewis(a) and sialyl-Lewis(x) antigens in colorectal adenocarcinoma. Histopathology 40:440–449

Biancone L, Araki M, Araki K, Vassalli P, Stamenkovic I (1996) Redirection of tumor metastasis by expression of E-selectin in vivo. J Exp Med 183:581–587

Bitler BG, Menzl I, Huerta CL, Sands B, Knowlton W, Chang A, Schroeder JA (2009) Intracellular MUC1 peptides inhibit cancer progression. Clin Cancer Res 15:100–109

Borsig L (2010) Antimetastatic activities of heparins and modified heparins. Experimental Evidence Thromb Res 125(Suppl 2):S66–S71

Borsig L, Wong R, Feramisco J, Nadeau DR, Varki NM, Varki A (2001) Heparin and cancer revisited: mechanistic connections involving platelets, P-selectin, carcinoma mucins, and tumor metastasis. Proc Natl Acad Sci USA 98:3352–3357

Borsig L, Wong R, Hynes RO, Varki NM, Varki A (2002) Synergistic effects of L- and P-selectin in facilitating tumor metastasis can involve non-mucin ligands and implicate leukocytes as enhancers of metastasis. Proc Natl Acad Sci USA 99:2193–2198

Borsig L, Stevenson JL, Varki A (2007) Heparin in cancer: role of selectin interactions. In: Khorana AA, Francis CW (eds) Cancer-associated thrombosis. Informa Healthcare, New York, pp 97–113

Brockhausen I (1999) Pathways of O-glycan biosynthesis in cancer cells. Biochim Biophys Acta 1473:67–95

Brockhausen I (2006) Mucin-type O-glycans in human colon and breast cancer: glycodynamics and functions. EMBO Rep 7:599–604

Brockhausen I, Yang JM, Burchell J, Whitehouse C, Taylor-Papadimitriou J (1995) Mechanisms underlying aberrant glycosylation of MUC1 mucin in breast cancer cells. Eur J Biochem 233:607–617

Brockhausen I, Yang J, Dickinson N, Ogata S, Itzkowitz SH (1998) Enzymatic basis for sialyl-Tn expression in human colon cancer cells. Glycoconj J 15:595–603

Brown JR, Fuster MM, Li R, Varki N, Glass CA, Esko JD (2006) A disaccharide-based inhibitor of glycosylation attenuates metastatic tumor cell dissemination. Clin Cancer Res 12:2894–2901

Burdick MD, Harris A, Reid CJ, Iwamura T, Hollingsworth MA (1997) Oligosaccharides expressed on MUC1 produced by pancreatic and colon tumor cell lines. J Biol Chem 272:24198–24202

Byrd JC, Bresalier RS (2004) Mucins and mucin binding proteins in colorectal cancer. Cancer Metastasis Rev 23:77–99

Chang SK, Dohrman AF, Basbaum CB, Ho SB, Tsuda T, Toribara NW, Gum JR, Kim YS (1994) Localization of mucin (MUC2 and MUC3) messenger RNA and peptide expression in human normal intestine and colon cancer. Gastroenterology 107:28–36

Chaturvedi P, Singh AP, Batra SK (2008) Structure, evolution, and biology of the MUC4 mucin. FASEB J 22:966–981

Chen L, Zhang W, Fregien N, Pierce M (1998) The her-2/neu oncogene stimulates the transcription of N-acetylglucosaminyltransferase V and expression of its cell surface oligosaccharide products. Oncogene 17:2087–2093

Chiang AC, Massague J (2008) Molecular basis of metastasis. N Engl J Med 359:2814–2823

Coussens LM, Werb Z (2002) Inflammation and cancer. Nature 420:860–867

Crocker PR, Paulson JC, Varki A (2007) Siglecs and their roles in the immune system. Nat Rev Immunol 7:255–266

Demetriou M, Nabi IR, Coppolino M, Dedhar S, Dennis JW (1995) Reduced contact-inhibition and substratum adhesion in epithelial cells expressing GlcNAc-transferase V. J Cell Biol 130:383–392

Dennis JW, Laferte S (1987) Tumor cell surface carbohydrate and the metastatic phenotype. Cancer Metastasis Rev 5:185–204

Dennis JW, Laferte S, Waghorne C, Breitman ML, Kerbel RS (1987) Beta 1-6 branching of Asn-linked oligosaccharides is directly associated with metastasis. Science 236:582–585

Dennis JW, Pawling J, Cheung P, Partridge E, Demetriou M (2002) UDP-N-acetylglucosamine: alpha-6-D-mannoside beta1, 6 N-acetylglucosaminyltransferase V (Mgat5) deficient mice. Biochim Biophys Acta 1573:414–422

Dennis JW, Lau KS, Demetriou M, Nabi IR (2009) Adaptive regulation at the cell surface by N-glycosylation. Traffic 10:1569–1578

Feizi T (1985) Demonstration by monoclonal antibodies that carbohydrate structures of glycoproteins and glycolipids are onco-developmental antigens. Nature 314:53–57

Freire T, Berois N, Sonora C, Varangot M, Barrios E, Osinaga E (2006) UDP-N-acetyl-D-galactosamine:polypeptide N-acetylgalactosaminyltransferase 6 (ppGalNAc-T6) mRNA as a potential new marker for detection of bone marrow-disseminated breast cancer cells. Int J Cancer 119:1383–1388

Fukuda M, Spooncer E, Oates JE, Dell A, Klock JC (1984) Structure of sialylated fucosyl lactosaminoglycan isolated from human granulocytes. J Biol Chem 259:10925–10935

Fuster MM, Esko JD (2005) The sweet and sour of cancer: glycans as novel therapeutic targets. Nat Rev Cancer 5:526–542

Gao J, McConnell MJ, Yu B, Li J, Balko JM, Black EP, Johnson JO, Lloyd MC, Altiok S, Haura EB (2009) MUC1 is a downstream target of STAT3 and regulates lung cancer cell survival and invasion. Int J Oncol 35:337–345

Girnita L, Wang M, Xie Y, Nilsson G, Dricu A, Wejde J, Larsson O (2000) Inhibition of N-linked glycosylation down-regulates insulin-like growth factor-1 receptor at the cell surface and kills Ewing's sarcoma cells: therapeutic implications. Anticancer Drug Des 15:67–72

Gomes J, Marcos NT, Berois N, Osinaga E, Magalhaes A, Pinto-de-Sousa J, Almeida R, Gartner F, Reis CA (2009) Expression of UDP-N-acetyl-D-galactosamine: polypeptide N-acetylgalactosaminyltransferase-6 in gastric mucosa, intestinal metaplasia, and gastric carcinoma. J Histochem Cytochem 57:79–86

Granovsky M, Fata J, Pawling J, Muller WJ, Khokha R, Dennis JW (2000) Suppression of tumor growth and metastasis in Mgat5-deficient mice. Nat Med 6:306–312

Guda K, Moinova H, He J, Jamison O, Ravi L, Natale L, Lutterbaugh J, Lawrence E, Lewis S, Willson JK, Lowe JB, Wiesner GL, Parmigiani G, Barnholtz-Sloan J, Dawson DW, Velculescu VE, Kinzler KW, Papadopoulos N, Vogelstein B, Willis J, Gerken TA, Markowitz SD (2009) Inactivating germ-line and somatic mutations in polypeptide N-acetylgalactosaminyltransferase 12 in human colon cancers. Proc Natl Acad Sci USA 106:12921–12925

Guo HB, Lee I, Kamar M, Akiyama SK, Pierce M (2002) Aberrant N-glycosylation of beta1 integrin causes reduced alpha5beta1 integrin clustering and stimulates cell migration. Cancer Res 62:6837–6845

Guo P, Wang QY, Guo HB, Shen ZH, Chen HL (2004) N-acetylglucosaminyltransferase V modifies the signaling pathway of epidermal growth factor receptor. Cell Mol Life Sci 61: 1795–1804

Hakomori S (1986) Tumor-associated glycolipid antigens, their metabolism and organization. Chem Phys Lipids 42:209–233

Hakomori S (2001) Tumor-associated carbohydrate antigens defining tumor malignancy: basis for development of anti-cancer vaccines. Adv Exp Med Biol 491:369–402

Hanisch FG (2001) O-glycosylation of the mucin type. Biol Chem 382:143–149

Hollingsworth MA, Swanson BJ (2004) Mucins in cancer: protection and control of the cell surface. Nat Rev Cancer 4:45–60

Hostettler N, Naggi A, Torri G, Casu B, Vlodavsky I, Borsig L (2007) P-selectin- and heparanase-dependent antimetastatic activity of non-anticoagulant heparins. FASEB J 21:3562–3572

Idikio HA (1997) Sialyl-Lewis-X, Gleason grade and stage in non-metastatic human prostate cancer. Glycoconj J 14:875–877

Iijima J, Zhao Y, Isaji T, Kameyama A, Nakaya S, Wang X, Ihara H, Cheng X, Nakagawa T, Miyoshi E, Kondo A, Narimatsu H, Taniguchi N, Gu J (2006) Cell-cell interaction-dependent regulation of N-acetylglucosaminyltransferase III and the bisected N-glycans in GE11 epithelial cells. Involvement of E-cadherin-mediated cell adhesion. J Biol Chem 281:13038–13046

Isaji T, Gu J, Nishiuchi R, Zhao Y, Takahashi M, Miyoshi E, Honke K, Sekiguchi K, Taniguchi N (2004) Introduction of bisecting GlcNAc into integrin alpha5beta1 reduces ligand binding and down-regulates cell adhesion and cell migration. J Biol Chem 279:19747–19754

Izumi Y, Taniuchi Y, Tsuji T, Smith CW, Nakamori S, Fidler IJ, Irimura T (1995) Characterization of human colon carcinoma variant cells selected for sialyl Lex carbohydrate antigen: liver colonization and adhesion to vascular endothelial cells. Exp Cell Res 216:215–221

Johansson ME, Phillipson M, Petersson J, Velcich A, Holm L, Hansson GC (2008) The inner of the two Muc2 mucin-dependent mucus layers in colon is devoid of bacteria. Proc Natl Acad Sci USA 105:15064–15069

Jorgensen T, Berner A, Kaalhus O, Tveter KJ, Danielsen HE, Bryne M (1995) Up-regulation of the oligosaccharide sialyl LewisX: a new prognostic parameter in metastatic prostate cancer. Cancer Res 55:1817–1819

Ju T, Lanneau GS, Gautam T, Wang Y, Xia B, Stowell SR, Willard MT, Wang W, Xia JY, Zuna RE, Laszik Z, Benbrook DM, Hanigan MH, Cummings RD (2008) Human tumor antigens Tn and sialyl Tn arise from mutations in Cosmc. Cancer Res 68:1636–1646

Julien S, Adriaenssens E, Ottenberg K, Furlan A, Courtand G, Vercoutter-Edouart AS, Hanisch FG, Delannoy P, Le Bourhis X (2006) ST6GalNAc I expression in MDA-MB-231 breast cancer cells greatly modifies their O-glycosylation pattern and enhances their tumourigenicity. Glycobiology 16:54–64

Kannagi R (1997) Carbohydrate-mediated cell adhesion involved in hematogenous metastasis of cancer. Glycoconj J 14:577–584

Kansas GS (1996) Selectins and their ligands: current concepts and controversies. Blood 88:3259–3287

Karin M, Greten FR (2005) NF-kappaB: linking inflammation and immunity to cancer development and progression. Nat Rev Immunol 5:749–759

Khatib AM, Kontogiannea M, Fallavollita L, Jamison B, Meterissian S, Brodt P (1999) Rapid induction of cytokine and E-selectin expression in the liver in response to metastatic tumor cells. Cancer Res 59:1356–1361

Kim YJ, Varki A (1997) Perspectives on the significance of altered glycosylation of glycoproteins in cancer. Glycoconj J 14:569–576

Kim YS, Gum J Jr, Brockhausen I (1996) Mucin glycoproteins in neoplasia. Glycoconj J 13: 693–707

Kim YJ, Borsig L, Varki NM, Varki A (1998) P-selectin deficiency attenuates tumor growth and metastasis. Proc Natl Acad Sci USA 95:9325–9330

Kobata A, Amano J (2005) Altered glycosylation of proteins produced by malignant cells, and application for the diagnosis and immunotherapy of tumours. Immunol Cell Biol 83:429–439

Kohlgraf KG, Gawron AJ, Higashi M, Meza JL, Burdick MD, Kitajima S, Kelly DL, Caffrey TC, Hollingsworth MA (2003) Contribution of the MUC1 tandem repeat and cytoplasmic tail to invasive and metastatic properties of a pancreatic cancer cell line. Cancer Res 63:5011–5020

Kornfeld R, Kornfeld S (1985) Assembly of asparagine-linked oligosaccharides. Annu Rev Biochem 54:631–664

Kragh M, Binderup L, Vig Hjarnaa PJ, Bramm E, Johansen KB, Frimundt Petersen C (2005) Non-anti-coagulant heparin inhibits metastasis but not primary tumor growth. Oncol Rep 14:99–104

Krause T, Turner GA (1999) Are selectins involved in metastasis? Clin Exp Metastasis 17: 183–192

Lau KS, Dennis JW (2008) N-Glycans in cancer progression. Glycobiology 18:750–760

Lau KS, Partridge EA, Grigorian A, Silvescu CI, Reinhold VN, Demetriou M, Dennis JW (2007) Complex N-glycan number and degree of branching cooperate to regulate cell proliferation and differentiation. Cell 129:123–134

Läubli H, Borsig L (2010a) Selectins as mediators of lung metastasis. Cancer Microenviron 3: 97–105

Läubli H, Borsig L (2010b) Selectins promote tumor metastasis. Semin Cancer Biol 20:169–177

Läubli H, Stevenson JL, Varki A, Varki NM, Borsig L (2006) L-selectin facilitation of metastasis involves temporal induction of fut7-dependent ligands at sites of tumor cell arrest. Cancer Res 66:1536–1542

Läubli H, Spanaus KS, Borsig L (2009) Selectin-mediated activation of endothelial cells induces expression of CCL5 and promotes metastasis through recruitment of monocytes. Blood 114:4583–4591

Leivonen M, Nordling S, Lundin J, von Boguslawski K, Haglund C (2001) STn and prognosis in breast cancer. Oncology 61:299–305

Li Y, Bharti A, Chen D, Gong J, Kufe D (1998) Interaction of glycogen synthase kinase 3beta with the DF3/MUC1 carcinoma-associated antigen and beta-catenin. Mol Cell Biol 18:7216–7224

Liu W, Ramachandran V, Kang J, Kishimoto TK, Cummings RD, McEver RP (1998) Identification of N-terminal residues on P-selectin glycoprotein ligand-1 required for binding to P-selectin. J Biol Chem 273:7078–7087

Ludwig RJ, Boehme B, Podda M, Henschler R, Jager E, Tandi C, Boehncke WH, Zollner TM, Kaufmann R, Gille J (2004) Endothelial P-selectin as a target of heparin action in experimental melanoma lung metastasis. Cancer Res 64:2743–2750

Mantovani A, Allavena P, Sica A, Balkwill F (2008) Cancer-related inflammation. Nature 454:436–444

Nakamori S, Kameyama M, Imaoka S, Furukawa H, Ishikawa O, Sasaki Y, Kabuto T, Iwanaga T, Matsushita Y, Irimura T (1993) Increased expression of sialyl Lewisx antigen correlates with poor survival in patients with colorectal carcinoma: clinicopathological and immunohistochemical study. Cancer Res 53:3632–3637

Nakamori S, Ota DM, Cleary KR, Shirotani K, Irimura T (1994) MUC1 mucin expression as a marker of progression and metastasis of human colorectal carcinoma. Gastroenterology 106:353–361

Nakamori S, Furukawa H, Hiratsuka M, Iwanaga T, Imaoka S, Ishikawa O, Kabuto T, Sasaki Y, Kameyama M, Ishiguro S, Irimura T (1997a) Expression of carbohydrate antigen sialyl Le(a): a new functional prognostic factor in gastric cancer. J Clin Oncol 15:816–825

Nakamori S, Kameyama M, Imaoka S, Furukawa H, Ishikawa O, Sasaki Y, Izumi Y, Irimura T (1997b) Involvement of carbohydrate antigen sialyl Lewis(x) in colorectal cancer metastasis. Dis Colon Rectum 40:420–431

Nierodzik ML, Karpatkin S (2006) Thrombin induces tumor growth, metastasis, and angiogenesis: evidence for a thrombin-regulated dormant tumor phenotype. Cancer Cell 10:355–362

Ogata S, Koganty R, Reddish M, Longenecker BM, Chen A, Perez C, Itzkowitz SH (1998) Different modes of sialyl-Tn expression during malignant transformation of human colonic mucosa. Glycoconj J 15:29–35

Ogawa JI, Inoue H, Koide S (1997) Alpha-2, 3-Sialyltransferase type 3N and alpha-1, 3-fucosyltransferase type VII are related to sialyl Lewis(x) synthesis and patient survival from lung carcinoma. Cancer 79:1678–1685

Okazaki K, Nakayama Y, Shibao K, Hirata K, Nagata N, Itoh H (1998) Enhancement of metastatic activity of colon cancer as influenced by expression of cell surface antigens. J Surg Res 78:78–84

Onder TT, Gupta PB, Mani SA, Yang J, Lander ES, Weinberg RA (2008) Loss of E-cadherin promotes metastasis via multiple downstream transcriptional pathways. Cancer Res 68: 3645–3654

Partridge EA, Le Roy C, Di Guglielmo GM, Pawling J, Cheung P, Granovsky M, Nabi IR, Wrana JL, Dennis JW (2004) Regulation of cytokine receptors by Golgi N-glycan processing and endocytosis. Science 306:120–124

Pinho S, Marcos NT, Ferreira B, Carvalho AS, Oliveira MJ, Santos-Silva F, Harduin-Lepers A, Reis CA (2007) Biological significance of cancer-associated sialyl-Tn antigen: modulation of malignant phenotype in gastric carcinoma cells. Cancer Lett 249:157–170

Renkonen J, Paavonen T, Renkonen R (1997) Endothelial and epithelial expression of sialyl Lewis (x) and sialyl Lewis(a) in lesions of breast carcinoma. Int J Cancer 74:296–300

Rhodes JM (1999) Usefulness of novel tumour markers. Ann Oncol 10(Suppl 4):118–121

Senapati S, Chaturvedi P, Sharma P, Venkatraman G, Meza JL, El-Rifai W, Roy HK, Batra SK (2008) Deregulation of MUC4 in gastric adenocarcinoma: potential pathobiological implication in poorly differentiated non-signet ring cell type gastric cancer. Br J Cancer 99:949–956

Shibao K, Izumi H, Nakayama Y, Ohta R, Nagata N, Nomoto M, Matsuo K, Yamada Y, Kitazato K, Itoh H, Kohno K (2002) Expression of UDP-N-acetyl-alpha-D-galactosamine-polypeptide galNAc N-acetylgalactosaminyl transferase-3 in relation to differentiation and prognosis in patients with colorectal carcinoma. Cancer 94:1939–1946

Shimodaira K, Nakayama J, Nakamura N, Hasebe O, Katsuyama T, Fukuda M (1997) Carcinoma-associated expression of core 2 beta-1, 6-N-acetylglucosaminyltransferase gene in human colorectal cancer: role of O-glycans in tumor progression. Cancer Res 57:5201–5206

Spicer AP, Rowse GJ, Lidner TK, Gendler SJ (1995) Delayed mammary tumor progression in Muc-1 null mice. J Biol Chem 270:30093–30101

Tang CK, Katsara M, Apostolopoulos V (2008) Strategies used for MUC1 immunotherapy: human clinical studies. Expert Rev Vaccines 7:963–975

Tarp MA, Clausen H (2008) Mucin-type O-glycosylation and its potential use in drug and vaccine development. Biochim Biophys Acta 1780:546–563

Tatsumi M, Watanabe A, Sawada H, Yamada Y, Shino Y, Nakano H (1998) Immunohistochemical expression of the sialyl Lewis x antigen on gastric cancer cells correlates with the presence of liver metastasis. Clin Exp Metastasis 16:743–750

Taylor-Papadimitriou J, Burchell J, Miles DW, Dalziel M (1999) MUC1 and cancer. Biochim Biophys Acta 1455:301–313

Ten Hagen KG, Fritz TA, Tabak LA (2003) All in the family: the UDP-GalNAc:polypeptide N-acetylgalactosaminyltransferases. Glycobiology 13:1R–16R

Tozawa K, Okamoto T, Kawai N, Hashimoto Y, Hayashi Y, Kohri K (2005) Positive correlation between sialyl Lewis X expression and pathologic findings in renal cell carcinoma. Kidney Int 67:1391–1396

Varki NM, Varki A (2002) Heparin inhibition of selectin-mediated interactions during the hematogenous phase of carcinoma metastasis: rationale for clinical studies in humans. Semin Thromb Hemost 28:53–66

Varki A, Cummings RD, Esko JD, Freeze HH, Stanley P, Bertozzi CR, Hart GW, Etzler ME (2008) Essentials of glycobiology. Cold Spring Harbor Laboratory Press, New York

Veiga SS, Chammas R, Cella N, Brentani RR (1995) Glycosylation of beta-1 integrins in B16-F10 mouse melanoma cells as determinant of differential binding and acquisition of biological activity. Int J Cancer 61:420–424

Velcich A, Yang W, Heyer J, Fragale A, Nicholas C, Viani S, Kucherlapati R, Lipkin M, Yang K, Augenlicht L (2002) Colorectal cancer in mice genetically deficient in the mucin Muc2. Science 295:1726–1729

Workman HC, Sweeney C, Carraway KL 3rd (2009) The membrane mucin Muc4 inhibits apoptosis induced by multiple insults via ErbB2-dependent and ErbB2-independent mechanisms. Cancer Res 69:2845–2852

Yamaguchi A, Ding K, Maehara M, Goi T, Nakagawara G (1998) Expression of nm23-H1 gene and Sialyl Lewis X antigen in breast cancer. Oncology 55:357–362

Yoshimura M, Nishikawa A, Ihara Y, Taniguchi S, Taniguchi N (1995) Suppression of lung metastasis of B16 mouse melanoma by N-acetylglucosaminyltransferase III gene transfection. Proc Natl Acad Sci USA 92:8754–8758

Zhao Y, Nakagawa T, Itoh S, Inamori K, Isaji T, Kariya Y, Kondo A, Miyoshi E, Miyazaki K, Kawasaki N, Taniguchi N, Gu J (2006) N-acetylglucosaminyltransferase III antagonizes the effect of N-acetylglucosaminyltransferase V on alpha3beta1 integrin-mediated cell migration. J Biol Chem 281:32122–32130

Zheng M, Fang H, Hakomori S (1994) Functional role of N-glycosylation in alpha 5 beta 1 integrin receptor. De-N-glycosylation induces dissociation or altered association of alpha 5 and beta 1 subunits and concomitant loss of fibronectin binding activity. J Biol Chem 269:12325–12331

Chapter 4
Glycosaminoglycans in Atherosclerosis and Thrombosis

C.P. Vicente, J.A.P. Godoy, and C.C. Werneck

Abstract The endothelium plays a crucial role in regulating vascular tone, inflammatory responses, thrombosis, and atherosclerosis. Part of these activities is regulated by the presence of glycosaminoglycans (GAGs) in the vessel wall and in the extracellular matrix (ECM) surrounding it. GAGs in atherosclerosis can help regulate atherogenesis through their ability to retain lipoproteins in the vessel wall. Prolonged retention of lipoproteins may render them susceptible to chemical modifications, leading to their aggregation, cellular uptake, and lipid accumulation. GAGs can also act as anticoagulants by interacting with proteins like antitrombin and heparin cofator II and thereby promoting their activation and increasing their ability to inhibit thrombin. In this chapter, we discuss the roles of different GAGs located in the vessel wall and ECM, focusing on understanding the mechanisms of action of these molecules in atherosclerosis and thrombosis.

4.1 The Vascular System

The vascular system is made up of vessels (arteries, arterioles, veins, venules, and capillaries) that carry blood throughout the body and deliver oxygen and nutrients to body tissues. All vessel lumens in vertebrates are lined with endothelial cells that rest on an underlying basement membrane – a specialized thin sheet-like structure composed of extracellular matrix (ECM) molecules such as laminins, type-IV collagens, nidogens, and perlecan. The medial layer is composed by vascular smooth muscle cells (VSMCs) that build up the vessels and are surrounded by a basement layer with similar organization as the endothelial basement membrane

C.P. Vicente (✉) and J.A.P. Godoy
Department of Anatomy, Cellular Biology, physiology and Biophysics, Biology Institute CP 6109, State University of Campinas – UNICAMP, Campinas, SP, Brazil
e-mail: cvicente@unicamp.br

C.C. Werneck
Department of Biochemistry, Biology Institute CP 6109, State University of Campinas – UNICAMP, Campinas, SP, Brazil

M.S.G. Pavão (ed.), *Glycans in Diseases and Therapeutics*, Biology of Extracellular Matrix, 83
DOI 10.1007/978-3-642-16833-8_4, © Springer-Verlag Berlin Heidelberg 2011

(Hallmann et al. 2005). In the media of large vessels, several layers of VSMCs are separated by elastic fibers and an ECM rich in collagens (Hultgardh-Nilsson and Durbeej 2007). The medial layer is covered by an outer layer, the adventitia, composed mainly of fibroblasts and collagen fibers (Fig. 4.1a). Most normal large- and medium-sized arteries contain large amounts of elastic fiber protein (as much as 50% of the total ECM protein), lesser amounts of collagen, and sparse amounts of proteoglycans and glycoproteins. In vascular diseases, these proportions change, with proteoglycans predominating in early vascular lesions and collagen found in abundance in later vascular lesions (Wight 2008). Proteoglycans and associated ECM molecules influence many of the events leading to the development of vascular lesions associated with lipid retention (Williams and Tabas 1995), calcification (Fischer et al. 2004), changes in smooth muscle cell (SMC) phenotypes (Evanko et al. 1999), and retention of macrophages (Wilkinson et al. 2006; Wight 2008). Alterations in one or more of these vessel components may lead to the loss of homeostasis of the vessel and consequently to vascular disease (Hultgardh-Nilsson and Durbeej 2007).

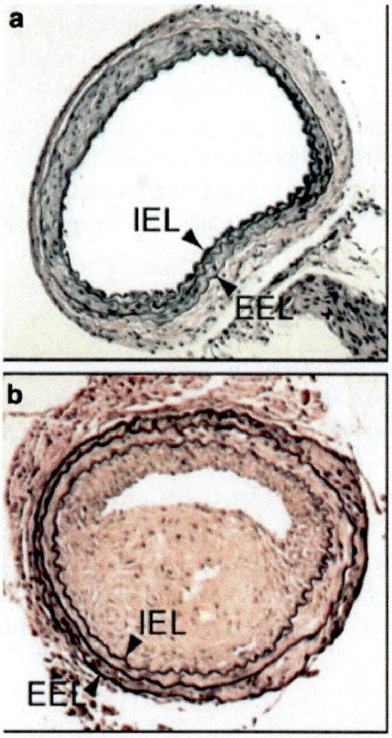

Fig. 4.1 (**a**) Cross-section of mouse carotid artery stained with Verhoff Van Gieson stain for elastic lamina (**b**) Cross section of mouse carotid artery stained with Verhoff Van Gieson stain for elastic lamina 21 days after arterial injury showing neointima occluding the lumen of the vessel (SMC – smooth muscle cells). This image was originally published in Vicente CP, He L, Tollefsen DM (2007) Accelerated atherogenesis and neointima formation in heparin cofactor II deficient mice. Blood 110:4261–4275. Published with permission

The endothelium functions as a permeability barrier, a thromboresistant surface, and as an initiator and/or inhibitor of numerous biologic processes, especially those related to inflammation, hemostasis and thrombosis, immunity, hemodynamic mechanotransduction, and vascular repair (Gotlieb 2005). Endothelial dysfunction is characterized by a shift of endothelial actions toward reduced vasodilation, a proinflammatory state, and development of prothrombotic properties. Endothelial dysfunction or damage also can cause growth of SMCs into the lumen of the vessel, forming a new layer of cells that can obstruct the vessel, called neointima (Fig. 4.1b). It is also associated with most forms of cardiovascular diseases, such as hypertension, coronary artery disease, chronic heart failure, peripheral artery disease, diabetes, and chronic renal failure. Mechanisms that participate in the reduction of vasodilatatory responses in endothelial dysfunction include reduced nitric oxide generation, oxidative stress, and reduced production of hyperpolarizing factor. Also, upregulation of adhesion molecules, generation of chemokines such as macrophage chemoattractant peptide-1, and production of plasminogen activator inhibitor-1 participate in the inflammatory response and contribute to a prothrombotic state that accelerates the establishment of atherosclerosis (Endemann and Schiffrin 2004).

4.2 Atherosclerosis

Atherosclerosis is the result of a complex interaction between lipoproteins, ECM, and cells of the vessel wall leading to the formation of a lesion known as atherosclerotic plaque. Hypertension, hypercholesterolemia, hypertriglyceridemia, hyperglycemia, modified low-density lipoproteins (LDLs), free radicals, infectious microorganisms, shear stress, toxins produced after smoking or combination of these, and other factors lead to a compensatory inflammatory response that can contribute to atherosclerosis. Current therapies target "risk factors" that contribute to increased incidence of cardiovascular disease (Little et al. 2007). The maximum efficacy of these strategies as determined from multiple clinical trials is limited to around 30% (Pignone et al. 2000). These risk factors promote endothelial dysfunction that can cause nitric oxide synthesis, promoting local oxidation of circulating lipoproteins and their entry into the vessel wall (Little et al. 2007). Intracellular reactive oxygen species similarly induced by the multiple atherosclerosis risk factors lead to enhanced oxidative stress in vascular cells and further activate intracellular signaling molecules involved in gene expression (Chatzizisis et al. 2007).

The pathogenesis of atherosclerosis is a complex multifactorial process. Shear stress induced by blood flow has emerged as an important factor in the pathogenesis of the focal fibroinflammatory lipid atherosclerotic plaque. It is one of the factors that can precipitate atherosclerosis and is a biomechanical force that is determined by blood flow, vessel geometry, and fluid viscosity. The parallel frictional drag force of shear stress is one of the most important blood flow-induced mechanical stresses acting to determine the structure and function of the vessel wall. Hemodynamic shear stress is considered an important local risk factor in atherogenesis.

Endothelial dysfunction and disturbed hemodynamic shear stress lead to erosion and fissure formation on the surface of complicated fibroinflammatory lipid plaques (Cunningham and Gotlieb 2005).

Inflammation is also a factor that plays an important role in the progression of atherosclerosis and plaque destabilization. Plaque destabilization can convert a chronic process into an acute disorder with ensuing thromboembolism and cardiac arrest. During atherosclerosis, T cells and macrophages infiltrate the vessel wall triggered by endothelial dysfunction and locally interact in a synergistic manner (Gotlieb 2005). When LDL is trapped in an arterial wall, it can undergo progressive oxidation and be internalized by macrophages through nonspecific surface scavenger receptors (Tabata et al. 2007). Autoreactive T cells recognize oxLDL, heparan, and chondroitin/dermatan sulfate proteoglycans and locally release proinflammatory cytokines. Macrophages stimulated by T cell-derived cytokines transform into foam cells after uptake of oxLDL and may secrete metalloproteinases (MMP) predisposing the plaques to subsequent rupture. On plaque rupture, tissue factor rich plaque material is placed in contact with the circulation and activates the extrinsic coagulation pathway (Stoll and Bendszus 2006) (Fig. 4.2).

Increased expression of cell adhesion molecules is also related to the initial onset of atherosclerosis, promoting inflammation and cell migration. It facilitates adherence of leukocytes to the dysfunctional endothelium and their subsequent transmigration into the vessel wall. These events can initiate the inflammatory reaction that promotes atherosclerosis. The reaction promoted by these agents is instrumental in the initiation of atherosclerotic plaques and their destabilization. Plaque destabilization causes arterial lesion and thrombosis and these factors can lead to restenosis. Restenosis is the result of neointima formation and remodeling after endothelial lesion. It is common after percutaneous coronary and peripheral interventions, and it decreases the long-term clinical success of these interventions (Witztum 1994a, b). Arterial lesion in atherosclerotic animals like ApoE-deficient mice not only leads to restenosis through neointima formation but also results in atherosclerotic plaques and accumulation of lipids in the lumen of the vessel (Fig. 4.3). Local therapies such as drug-eluting stents have successfully reduced the rate of balloon- or stent-

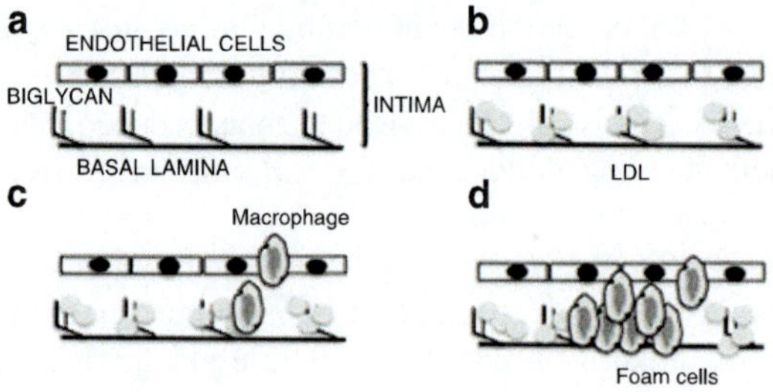

Fig. 4.2 Scheme of early atherosclerosis. (**a**) Normal intimal layer showing endothelial cells, basal lamina, and the presence of the Chondroitin/dermatan sulfate (CS/DS) proteoglycan biglycan. (**b**) Lipoprotein retention (LDL) in the CS/DS layer of biglycan. (**c**) Inflammatory response to lipid accumulation with migration of macrophages to the subendothelial layer. (**d**) Accumulation of macrophages, phagocytosis of the lipids, and formation of the foam cells

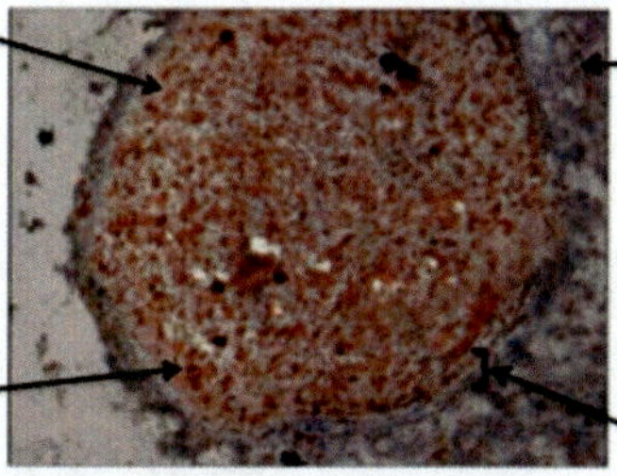

Fig. 4.3 Cross-section of mouse carotid artery stained with oil red and Gills hematoxylin. Observe the lumen of the vessel occluded with neointima (smooth muscle cells) and lipids

induced restenosis. The major areas of research that may provide a vascular-directed agent for the prevention of atherosclerosis are: inflammation (Ross 1999; Libby 2006), oxidation (Witztum 1994a, b; Stocker and Keaney 2004), endothelial dysfunction (Davignon 2004), diabetes-specific factors like hyperglycemia and advanced glycation end products (AGEs) (Brownlee 2001; Forbes et al. 2004), and lipid retention by vascular matrix specifically proteoglycans (Skalen et al. 2002; Little et al. 2007).

4.3 Diabetes and Hyperglycemia

Hyperglycemia induces modifications in arterial proteoglycans that may promote atherosclerosis. A positive relationship was observed between total glycosaminoglycans (GAGs) and cholesterol accumulation in diabetic but not in control arteries. In human arteries, the content of sulfated GAG has been shown to increase in early lesions but decrease with lesion development (Stevens et al. 1976). More importantly atherosclerosis-associated changes in specific GAG have been reported. The most significant change observed was a decrease in the ratio of heparan sulfate (HS): dermatan sulfate (DS) with atherosclerosis in both nondiabetic and diabetic tissues. Comparison of the diabetic and nondiabetic groups showed no difference in total GAG, but there is a significant change in distribution: an increase in DS and a decrease in HS in the diabetic tissues. These data indicated that atherosclerosis-associated changes in GAG were exacerbated in the diabetic state (Edwards et al. 2004).

4.4 Glycosaminoglycans and Proteoglycans in Atherosclerosis

GAGs are long, linear, and heterogeneous polysaccharides composed of repetitive disaccharide units containing an uronic acid (D-glucuronic acid or L-iduronic acid) and an amino sugar (D-glucosamine or D-galactosamine). The main GAGs are heparan sulfate (HS), dermatan sulfate, and chondroitin sulfate (CS) and they are composed of different disaccharide units as shown in Fig. 4.4. These molecules play an essential role in many biological functions, including growth control, signal

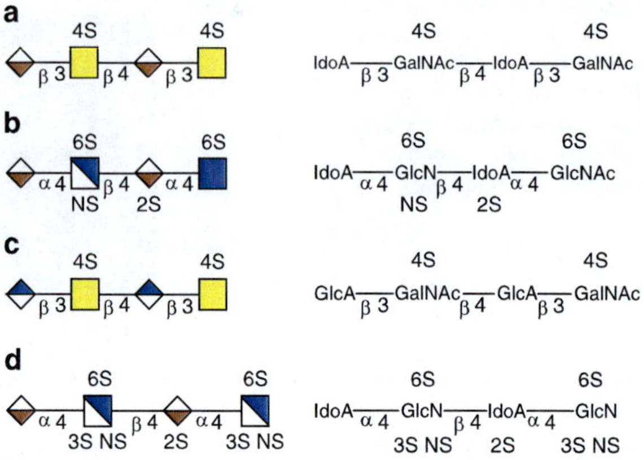

Fig. 4.4 Disaccharide composition of (**a**) dermatan sulfate, (**b**) heparan sulfate (**c**) chondroitin sulfate and (**d**) heparin. IdoA, L-iduronic acid. GalNAc, *N*-acetyl-D-galactosamine. GlcA, D-glucuronic acid. GlcNAc, *N*-acetyl-D-glucosamine. GlcN, glucosamine. 2S, 2-*O*-sulfate. 3S, 3-*O*-sulfate. 4S, 4-*O*-sulfate. NS, N-sulfated. The figure was created using GlycanBuilder (Ceroni et al. 2007)

transduction, cell adhesion, hemostasis, and lipid metabolism (Tumova et al. 2000; Volpi 2006). GAGs play a critical role in the assembly of protein–protein complexes such as growth factor receptors or enzyme inhibitors on the cell surface and in the ECM that are directly involved in initiating cell signaling events or inhibiting biochemical pathways. GAGs also participate in pathological processes, such as inflammation (Gozzo et al. 2002), microbial pathogenesis, and cancer (Monzavi-Karbassi et al. 2007). Nevertheless, despite these vital functions, our current level of knowledge is extremely limited since GAGs seem to perform several other important functions that are so far unknown.

Proteoglycans (PG) are highly charged complex molecules consisting of a core protein to which long chains of negatively charged GAGs are attached. There are four major classes of proteoglycans in the vascular system: heparan sulfate proteoglycan (HSPG) mainly represented by perlecan, chondroitin sulfate proteoglycan (CSPG) represented by versican, dermatan sulfate proteoglycan (DS) represented by decorin, and CS/DSPG represented by biglycan. The role of proteoglycans in atherosclerosis has been the subject of intense investigation largely owing to the observation that CS/DS proteoglycans associate with apoB-containing lipoproteins in vitro (Hurt-Camejo et al. 1997; Camejo et al. 2002). Chondroitin and dermatan sulfate (CS/DS) proteoglycans, like versican and biglycan, are considered proatherogenic because of their ability to retain LDLs (Wight and Merrilees 2004). By contrast, heparan sulfate (HS) proteoglycans have been proposed to be antiatherogenic, because decreased heparan sulfate is associated with increased atherosclerosis in different species (Tran-Lundmark et al. 2008).

One hypothesis for the role of proteoglycans in atherosclerosis is the retention hypothesis. It proposes that one of the earliest steps in the formation of an atherosclerotic plaque is the retention of atherogenic lipoproteins within the subendothelial space by their interactions with vascular proteoglycans (Williams and Tabas 1995). Proteoglycans bind lipoproteins through ionic interactions between the negatively charged sulfate and carboxyl groups on the GAG chains and the positively charged amino acid residues on apolipoproteins (apo) B and E. The critical role of lipoprotein retention by artery wall proteoglycans in the initiation of atherosclerosis was shown in a series of elegant experiments in which mice expressing proteoglycan binding-defective lipoproteins developed significantly less atherosclerosis than their littermates expressing either control lipoproteins or LDL receptor-binding-defective lipoproteins, despite similar levels of cholesterol (Skalen et al. 2002). It was demonstrated that increase in vascular proteoglycans content precede and contribute to atherosclerosis development (Huang et al. 2008). Thus, lipoprotein retention by vascular proteoglycans is a fundamental step in the initiation of atherosclerosis. Also, lipid retention is mainly associated with increased expression of the chondroitin/dermatan sulfate proteoglycan, biglycan. Apolipoproteins bind to the GAG chains of proteoglycans. It was observed that structural changes in GAG chains occur during the development of endothelial dysfunction, these changes may enhance LDL binding to these molecules which might be the critical initiating event in atherosclerosis (Wilson et al. 2008) (Fig. 4.5).

Several molecules are described as being able to alter GAGs length, sulfation pattern, and/or synthesis. One example is endothelin-1, a vascular growth factor and vasoconstrictor peptide. It is able to influence vascular structure and function through signaling directly via specific 7 transmembrane G protein-couple receptors and indirectly through transactivation of epidermal growth factor receptor. Endothelin-1

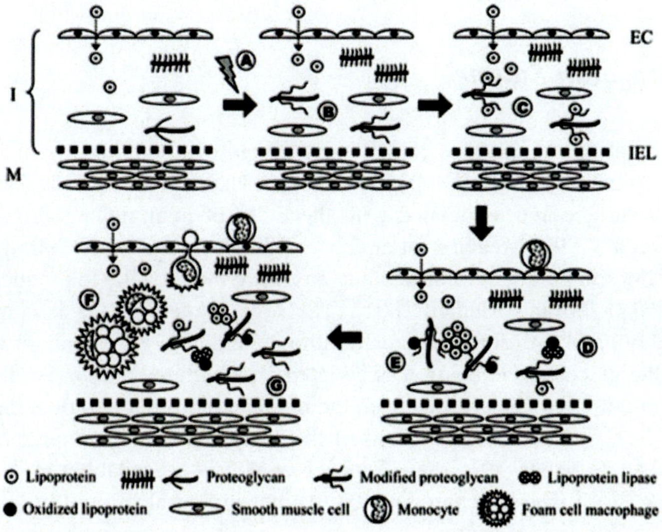

Fig. 4.5 Scheme of response-to-retention hypothesis (**a**) predisposing stimuli (**b**) direct binding of lipoproteins and proteoglycans with accessory molecules (**c**) direct binding of lipoprotein and proteoglycans (**d**) indirect binding of lipoprotein and proteoglycan with acessory proteins (**e**) modifications of lipoproteins (**f**) taking up lipid proteoglycans complexes by macrophages and (**g**) production of proteoglycans with a high binding affinity to lipoproteins. *I* intima, *M* media, *E* endothelial cells, *IEL* internal elastic lamina. This research was originally published in Nakashima Y, Wight TN, Sueishi K. Early atherosclerosis in humans: role of diffused intimas thickening and extracellular proteoglycans. Cardiovasc Res 79:14–23. Published with permission

stimulates proteoglycan synthesis, specifically stimulating the synthesis of longer GAG chains with an increased 6 to 4 sulfation position ratio of the disaccharides. Elongation of GAG chains is associated with greater binding of LDL. The 6-sulfate group of a GAG chain disaccharide is on a noncyclic carbon and is more accessible to bind to the ApoB100 moiety on LDL (Ballinger et al. 2004; Little et al. 2007). An increased amount of sulfate groups at the six positions on the *N*-acetyl-galactosamine residue of the disaccharide is associated with increased binding to LDL and greater atherogenicity (Cardoso and Mourao 1994). The effect of endothelin-1 is mediated via the ETA (endothelin A) receptor and further downstream signaling is via protein kinase C (PKC). Endothelin-1 effects were specific for stimulation of GAG synthesis since there was no effect on proteoglycan core protein synthesis, although there was an effect on total protein synthesis consistent with the mitogenic action of endothelin-1 on VSMC (Bobik et al. 1990). Oxidized LDL, free fatty acids, androgens, TGF-β (Transforming growth factor β), angiotensin II, and thrombin stimulate the synthesis of proteoglycans in VSMCs with longer GAG chains (Ivey and Little 2008; Little et al. 2002; Chang et al. 2000; Hashimura et al. 2005). Although the identity of multiple enzymes involved in GAG synthesis and elongation are well known, little is known about their regulation in VSMC. Further studies are necessary to help understand this complex mechanism that is directly involved in the initiation of atherosclerosis (Ballinger et al. 2009).

4.4.1 Heparan Sulfate

Heparan sulfate proteoglycans (HSPG) are ubiquitous components of all tissue-organized animal life forms (Dietrich et al. 1998; Nader et al. 2004) and are present on the cell surface and associated with the ECM of mammalian cells in culture (Bernfield et al. 1999; Medeiros et al. 2000; Whitelock and Iozzo 2005). Virtually all cells, from simple invertebrates to humans, have the capacity to produce heparan sulfate (HS) (Esko and Lindahl 2001). The sugar moiety of HS is composed of alternating units of a-D-glucosamine and uronic acid (b-D-glucuronic or a-L-iduronic) with the glucosamine being *N*-acetylated or *N*-sulfated and/or 6-sulfated. The fine structure of HS chains results from the regulated introduction of sulfate groups at the *N*-, 2-*O*, 6-*O*, and 3-*O* positions of the sugar chain and epimerization of the glucuronic acid (Nader et al. 1987; Dietrich et al. 1998; Lindahl et al. 1998; Esko and Selleck 2002). The HS chains exhibit distinct structural variability according to the tissue and species of origin.

The proportion of HSPG to total proteoglycan is approximately 41% in areas of normal human aorta and is found to be reduced to 20% in atherosclerotic areas. In human coronary arteries, a decrease in HS content and a proportional increase in dermatan and chondroitin sulfate are associated with increased atherosclerosis (Murata and Yokoyama 1982). The HSPG content is also markedly reduced in atherosclerotic human cerebral arteries and in atherosclerotic lesions from diabetic monkeys (Yoshino et al. 1997; Edwards et al. 2004; Tran et al. 2007).

Perlecan is the predominant HS proteoglycan in the arterial ECM, and it is found in basement membranes underneath the endothelial layer and around medial SMCs (Iozzo et al. 1994). Perlecan consists of a core protein with three attachment sites for HS GAGs side chains in domain I (Noonan et al. 1991; Kallunki and Tryggvason 1992) and an additional site in domain V. It was demonstrated that perlecan is downegulated in human carotid atherosclerosis (Tran et al. 2007) on the other hand, together with biglycan, they have been shown to be the predominant proteoglycans in atherosclerotic lesions in mice demonstrating that HS role in atherogenesis might be species specific. The deletion of the perlecan gene causes embryonic lethality with severe cephalic, myocardial, and cartilage abnormalities, including transposition of the vessels. These animals also present abnormally abundant mesenchymal cells expressing SMCs specific actin-isoform in the left ventricular outflow tract, which can lead to varying levels of the obstruction of the tract (Segev et al. 2004). Mice with complete perlecan deficiency are embryonic lethal due to the rupture of basal membranes in the heart and abnormal cartilage development. It was demonstrated that HS chains in perlecan played an important role in the regulation of wound healing, tumor growth, and angiogenesis. This result is most likely related to the ability of HS from perlecan to bind FGF-2 (fibroblast growth factor-2) and other growth factors, protecting them from proteolytic degradation and sequestering them in a latent state (Zhou et al. 2004) (Table 4.1).

Antiatherogenic properties of HS may involve several mechanisms. Enzymatic removal of HS has been shown to increase binding of LDL to endothelial matrix in vitro (Pillarisetti et al. 1997), indicating that HS may interfere with lipoprotein retention (Fig. 4.6). HS also reduces endothelial permeability for LDL, (Duan et al. 2005; Guretzki et al. 1994) and removal of HS chains has been shown to increase monocyte binding to the vessel wall. In addition, heparin and HS are potent inhibitors of SMC proliferation, which may influence plaque stability and size (Tran-Lundmark et al. 2008). Perlecan is present in the lesions of apoE- and LDL receptor-deficient mice. It was detected in intermediate and advanced lesions of hypercholesterolemic no-human primates and in cultures of medial SMCs from human atherosclerotic tissue, but its expression in human lesions has not yet been reported (Edwards et al. 2004; Tran et al. 2007; Iozzo et al. 1994). The LDL binding domain in perlecan shares homology with the binding domain of the LDL receptor and could potentially bind lipoproteins (Kreuger et al. 2002). Perlecan can bind growth factors that promote angiogenic and tumor growth functions. In the arterial wall, perlecan may inhibit the proliferation of SMCs as it does in cell culture (Tran et al. 2004). It also helps promoting atherosclerosis in mice, most likely through

Table 4.1 Possible roles of perlecan in atherogenesis

Lipoprotein retention	Inhibits binding of lipoproteins to fibronectin
Monocyte recruitment	Inhibits monocyte binding to subendothelial matrix by inhibiting monocyte cell surface HSPG interaction with matrix
Macrophage differentiation	Sequestry of cytokines, regulates their bioavaiability
Smooth muscle cell proliferation	Inhibits smooth muscle cell proliferation
Thrombosis	Potentiation of antithrombin inhibition of thrombin activity

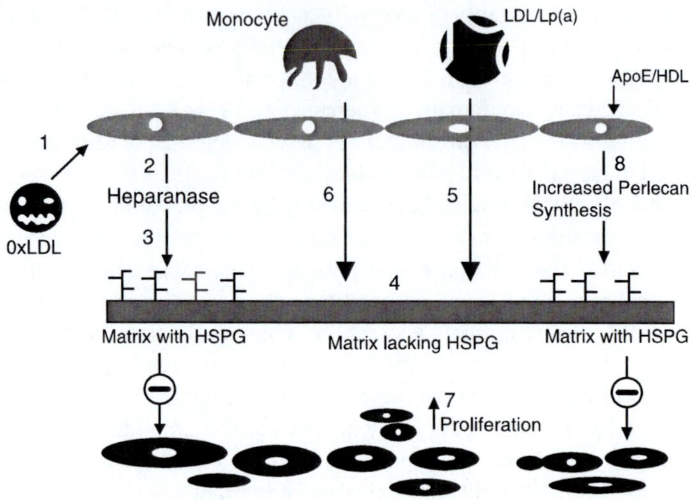

Fig. 4.6 Subendothelial matrix-associated heparan sulfate proteoglycan is atheroprotective. Subendothelial matrix contains adhesion proteins and heparan sulfate proteoglycans (HSPG). The major matrix HSPG is perlecan. Existing data suggest that matrix HSPG inhibits events related to atherogenesis. Atherosclerotic vessels have reduced HSPG, and in vitro, a decrease in HSPG is observed when endothelial cells are exposed to oxidized LDL (OxLDL) (1). This decrease in matrix HSPG could be due to decreased perlecan expression or induced secretion of a HS-degrading heparanase (2). Heparanase-mediated degradation of HS (3) leads to generation of matrix devoid of HS (4). Removal of HSPG may not only affect endothelial barrier function but could also increase retention of atherogenic lipoproteins [LDL and Lp(a)] (5) and circulating monocytes (6). Loss of HS may also promote SMC to migrate and proliferate in the subendothelial intima (7). ApoE and ApoE HDL may counter this by increasing the expression of perlecan (8) as well as by increasing the sulfation of HS. This research was originally published in Pillarisetti S (2000) Lipoprotein modulation of subendothelial heparan sulfate proteoglycans (perlecan) and atherogenesis. Trends Cardiovasc Med 10:60–65

increased retention of lipoproteins. In addition, the ability of HS to regulate SMC proliferation was found to influence SMC content in lesions, thus suggesting a role for perlecan in plaque stability. Because of differences in proteoglycans expressed in mice and humans, it is difficult to determine a role for perlecan in human disease (Tran-Lundmark et al. 2008).

4.4.2 Chondroitin/dermatan Sulfate Proteoglycans

Structural studies of Chondroitin sulfate (CS) chains from vertebrate and especially mammalian species have shown the extreme heterogeneity resulting from the organization of the repeating disaccharide units. Mounting evidence shows that the structure of CS depends on the animal species, tissue, physiological or pathological conditions, and the biological source (Malavaki et al. 2008).

Versican, a large chondroitin sulfate proteoglycan (CSPG), and biglycan, a small leucine rich dermatan sulfate PG (CS/DS), have been shown to be responsible for LDL accumulation in atherosclerotic vascular walls (Wight and Merrilees 2004). Typically, CS chains attached to the core protein of versican differ in size and composition depending on the tissue source or culture conditions. They vary in size from 25 to 80 kDa and contain different ratios of chondroitin 6-sulfate to chondroitin 4-sulfate. Their length and composition are regulated and controlled by different stimuli. Versican can be produced by SMC and has the ability to bind both hyaluronan and proteins in the fibulin family. This network, built up by versican, hyaluronan, and fibulins, can be important to SMC function during vessel injury, altering the proliferation and migration of these cells to form neointima (Wight 2002, 2008).

Versican is prominent in the intima and adventitia of most arteries and veins. It is likely that accumulation of versican in the intima is responsible for the proteoglycan-rich nature of this layer. Versican is present in early intimal thickening in what characterizes developing atherosclerosis and is associated with arterial smooth muscle cells. Versican is markedly increased in stented and nonstented restenotic lesion in humans. Hyaluronan also accumulates in versican-rich regions of arterial lesions. Clearly, versican is in the center of many key processes in vascular pathology. Elevated levels of versican in the vessel intima shift the balance in favor of pathological changes and contribute to increased intimal thickness, cell migration, proliferation, and adhesion of inflammatory cells (Wight and Merrilees 2004).

The main dermatan sulfate proteoglycans are decorin and biglycan. Decorin and biglycan contain 1 and 1-2 DS chains, respectively. DS is composed of linear polysaccharides assembled as disaccharide units containing L-iduronic acid (idoA) and *N*-acetyl-D-galactosamine (GalNAc), with 2-*O* sulfation of IdoA and sulfation of 4-*O* and/or 6-*O* of GalNAc joined by β 1,4 or β 1,3 linkages, respectively (Trowbridge and Gallo 2002).

Mice with a targeted disruption of the decorin gene are viable but have fragile skin with markedly reduced tensile strength. Ultrastructural analysis revealed abnormal collagen morphology in skin and tendon with coarser and irregular fiber outlines. These animals present uncontrolled lateral fusion of collagen fibrils, which accounts for the reduced tensile strength observed in the skin. These findings demonstrate a fundamental role for decorin in regulating collagen fiber formation in vivo (Danielson et al. 1997). Although the mechanism underlying the effects of decorin in atherosclerosis is not completely understood, it has been shown that it suppresses atherosclerosis progression. Decorin inhibits TGF-β, a cytokine that promotes atherosclerosis, by forming an inactive decorin core protein–TGF-β complex (Yamaguchi et al. 1990). Dermatan sulfate chains bound to decorin core proteins exhibit an anticoagulant activity by activating heparin cofactor II (Whinna et al. 1993). Furthermore, decorin interacts with a variety of proteins that are involved in matrix assembly (Thieszen and Rosenquist 1995). Cellular events influenced by decorin may affect atherosclerosis progression. Decorin is also an important modulator of lipid accumulation in collagen-rich regions of the arterial intima. It can bind lipoprotein lipase and phospholipase A2, which may explain the preferential accumulation of LDL in the vicinity of collagen fibers in the arterial intima. The role of decorin in atherogenesis may extend beyond its potential role in lipid accumulation. Decorin can bind fibronectin (Schmidt et al. 1991; Winnemoller et al. 1991), thrombospondin (Winnemoller et al. 1992), the C1q component of complement (Krumdieck et al. 1992), and TGF-β (Yamaguchi et al. 1990), each of which has been suggested to play a role in the development of atherosclerosis. Decorin is also associated with calcium deposits in lesions from patients undergoing sudden death suggesting that the current findings in cultured SMCs might be of relevance for calcification of human plaques. In human atherosclerosis, macrophages colocalize with both decorin (Evanko et al. 1998) and calcium deposits (Jeziorska et al. 1998). Because macrophages release IL-1, which induces decorin expression in vascular SMC (Edwards et al. 1994), it is conceivable that macrophages may cause increased calcification of atherosclerotic plaques by inducing decorin expression. Decorin can also be induced by high glucose concentrations in cultured mesangial cells and in mouse models of type 1 diabetes (Wahab et al. 2000), so, it seems possible that decorin plays a role in the severe vascular calcification occurring in diabetic patients (Kovanen and Pentikainen 1999; Fischer et al. 2004).

Biglycan-deficient mice (*bgn–/0*) are viable and have no profound skeletal patterning abnormalities at birth, however, with age, X-ray images showed that long bones had decreased length coupled with decreased mineral density and mass compared to age matched controls. Histological analysis showed that the mice have less trabecular volume and reduced cortical thickness (Xu et al. 1998). The thinned cortices in the absence of biglycan may explain why the bones also had reduced biomechanical strength. Bones from the *bgn−/0* mice have a reduced mineralization (Corsi et al. 2002). The response of *bgn−/0* cells to TGF-β is lower than in normal cells. In some tissues, biglycan could enhance TGF-β function (as in stromal cells), while in other cases (such as in fibrosis) it would inhibit or buffer TGF-β activities. It was also observed that in biglycan-deficient mice, the teeth had broader, more porous dentin, thicker enamel, and a mantle dentin mineralization near the dentin–enamel junction. Muscles from the *bgn−/0* were also affected, a defect presumed to be related to the absence of biglycan for binding to one of the members of the muscle plasma membrane, alpha dystroglycan (Bowe et al. 2000). Little et al. (2008) suggests that a small amount of biglycan with "natural" GAG chains typically present in the vessel wall is not proatherogenic. However, changes in the vessel wall associated with the expression of growth factors or hormones and a phenotypic "switch" in the resident intimal VSMCs leads to increased expression of biglycan and elongation of their GAG chains that, together, greatly increase the retention of lipoproteins and trigger the atherosclerotic cascade.

Adiponectin is a unique regulatory factor for PG synthesis in SMCs. Adiponectin is an adipocyte-derived plasma protein that exhibits antiatherogenic activity. It regulates decorin core protein synthesis and also biglycan/decorin sulfate GAG formation. The synthesis of decorin along with the inhibition of GlcA-to-IdoA epimerization in dermatan sulfate chains may influence the antiatherogenic activity of adiponectin in apolipoprotein E-deficient mice (ApoE −/−), where adiponectin reduces atherosclerotic changes such as intimal thickening (Okamoto et al. 2002). It is likely that the inhibition of VSMC proliferation is important for adiponectin's antiatherogenic activity (Matsuda et al. 2002; Yao et al. 2008).

4.4.3 Hyaluronic Acid

Hyaluronan (HA) is a high molecular mass (1×10^5–1×10^7 Da) unbranched GAG, composed of repeating disaccharides of *N*-acetyl-D-glucosamine and D-glucuronic acid (Laurent and Fraser 1992). It is synthesized at the cytoplasmic surface of the plasma membrane by a family of HA synthases (HASs) (Philipson and Schwartz 1984; Prehm 1984), and it is subsequently extruded through the plasma membrane into the pericellular matrix (Prehm 1984). HAS enzymes are encoded by three highly conserved mammalian genes, *HAS1*, *HAS2*, and *HAS3*, and are predicted integral plasma membrane proteins with multiple transmembrane domains (Itano et al. 1999; Weigel et al. 1997) that partially colocalize with membrane lipid microdomains (Kultti et al. 2006). HA degradation occurs locally through at least two catabolic pathways as follows: internalization by an endocytic pathway that requires functional CD44 (Knudson et al. 2002), a membrane microdomain-associated protein (Ilangumaran et al. 1998) and the principal cell-surface receptor for HA (Aruffo et al. 1990); and an extracellular catalysis mediated by hyaluronidase-2, a glycerophosphoinositol-anchored plasma membrane protein (Stern 2003). The half-life of HA in tissues ranges from a few hours to several days. Interstitial HA is catabolized locally or in the lymphatics vessels and by endothelial cells of the liver sinusoids (Laurent and Fraser 1992; Fraser et al. 1997). Thus, in tissues where interstitial fluid normally drains into the lymphatic system, a rapid rate of turnover by local cells is not necessary to avoid excessive accumulation of HA. Considerable work indicates that HA, a widely distributed component of the vertebrate tissues ECM accumulates in human and animals atherosclerotic lesions. (Evanko et al. 1998; Cuff et al. 2001). Furthermore, *HAS2* transgene overexpression promotes atherosclerosis in apolipoprotein E knockout mice (Chai et al. 2005). HA is frequently found in lipid-enriched areas of atherosclerotic lesions. It also coisolates with lipoproteins from human atherosclerotic lesions (Srinivasan et al. 1980) and promotes foam cell and cholesterol accumulation in skin xanthoma (Seike et al. 2006). It was demonstrated that in a rabbit model using LDL receptor-deficient Watanabe heritable hyperlipidemic (WHHL) rabbits, arterial smooth muscle cells (ASMCs) accumulated two- to fourfold more HA than normo-lipidemic ASMC. In addition, the HA synthesized by WHHL ASMC have a larger average hydrodynamic size. These changes appear to be due to a combination of increased synthesis and decreased degradation of HA in WHHL ASMC. These findings directly link cholesterol homeostasis in cells to the HA metabolism and that may help to explain HA accumulation in atherosclerotic lesions (Sakr et al. 2008).

4.5 Glycosaminoglycans and Thrombosis

Hemostasis is a defense against uncontrolled hemorrhage. But, if this physiological process occurs in an uncontrolled way, it can occlude blood vessels with thrombi and lead to events, such as unstable angina and myocardial infarction (Jennings 2009).

Arterial and venous thromboses, with their clinical manifestations such as stroke, myocardial infarction (MI), or pulmonary embolism, are the major causes of death in developed countries (Endler et al. 2002).

Abnormalities in endothelial function (vessel wall), blood flow, or blood components, either alone or in combination, can promote thrombus formation. Among the blood components are included abnormalities in platelet function, coagulation, and fibrinolysis. The integrity of the vessel wall, with its inner lining of endothelium, is crucial to the maintenance of a functional vasculature. The endothelium contains three main thromboregulators: nitric oxide, (Ignarro et al. 1987; Palmer et al. 1987), prostacyclin, (Marcus et al. 2002) and the ectonucleotidase CD394, which together, provide a defense against thrombus formation.

Thrombus formation is dependent on activation or exposure of these factors: collagen, tissue factor, factor V, platelets thrombin, and fibrin. Collagen and tissue factor that are present at the subendothelial matrix are important for the initiation of thrombus after vessel injury. When a lesion in the endothelial wall occurs, collagen and tissue factors are exposed to the flowing blood, initiating the coagulation cascade. Platelets, when exposed to tissue factor, initiate formation of thrombin what converts fibrinogen to fibrin and is also able to activate platelets. Thrombin also activates factor XIII into XIIIa, which helps associating fibrin to form a stabilized matrix, generating the fibrin clot. The fibrin clot containing fibronectin has an important role in hemostasis. The association of fibrin with a matrix protein like fibronectin serves as a provisional matrix supporting the migration and proliferation of inflammatory cells and fibroblasts (Dugan et al. 2006). Platelet adhesion and aggregation at sites of vascular injury are two main events in the control of hemostasis and development of thrombus. Platelets limit blood loss after injury and interfere in pathological conditions like deep vein thrombosis and arterial thrombosis (Furie and Furie 2008). Platelet thrombi have devastating effects in cardiovascular and cerebrovascular disease and they may be associated with 40% of all deaths in the USA (Denis and Wagner 2007). Factors that control thrombosis can be either prothrombotic or antithrombotic. The endothelium normally shows prothrombotic molecules like von Willebrand factor, factor V, tissue factor, plasminogen activator inhibitor-1, and antithrombotic molecules like prostacyclin (prostaglandin-I_2), endothelium-derived relaxing factor (EDRF), Ecto-ADPase, thrombomodulin, Tissue-type plasminogen activator (t-PA), cell-surface receptors for plasminogen, t-PA, urokinase-type plasminogen activator (u-PA), HSPG, and dermatan sulfate proteoglycans (DSPG) (Greiling and Clark 1997; Clark 2001; Tovar et al. 2005; Dugan et al. 2006). There are also factors like antithrombin (AT) and heparin cofactor II (HCII) proteins that circulate in plasma in inactive form and became active after endothelial lesion. These two molecules from the serpin family (serine protease inhibitors) are produced in liver. Their binding to GAGs such as heparan sulfate or dermatan sulfate makes them active, promoting their binding to thrombin, consequently inhibiting thrombosis.

Thrombosis may occur in arteries and in veins. Arterial walls are generally thicker than their venous counterparts. Veins have larger overall diameters, larger lumens, and thinner walls than their corresponding arteries (Gordillo and Sen 2003). Vascular diseases primarily affect arteries, and the most prevalent and clinically significant of these diseases is atherosclerosis, as discussed previously. Venous thromboembolic events are also important and may be related to congenital deficiencies of antithrombin, protein C, and protein S and are also frequently associated with neoplasias, immobility, pregnancy, or surgical procedures (Sen 2003; de Mattos et al. 2008). Some of these factors can cause deep vein thrombosis (DVT) and pulmonary embolism. Hematologic diseases associated with an increased risk of DVT (deep vein thrombosis) include disseminated intravascular coagulation (DIC), heparin-induced thrombocytopenia (HIT), antiphospholipid antibody syndrome, thrombotic thrombocytopenic purpura (TTP), hemolytic uremic syndrome (HUS), and myeloproliferative disorders (polycythemia vera and essential thrombocythemia) (Wakefield et al. 2008).

4.5.1 Heparin

Heparin is one of the most frequently prescribed medicines in the US. It is considered the preferential treatment for thromboembolism, surgical anticoagulation, and interventional cardiology (Kelton 2002). Heparin is a carbohydrate belonging to the family of the sulfated GAGs and is composed of repeating disaccharide units of 1 → 4 linked glucosamine and glucuronic acid/iduronic acid that contain sulfo-groups with a mean molecular weight of 15,000 (Liu and Pedersen 2007).

Heparin is able to bind antithrombin, factor Xa, Heparin Cofator II (HCII), and increases around 1,000× their ability to inhibit thrombin. In order to catalyze thrombin inhibition, heparin must bind both thrombin and antithrombin bridging them together. This reaction can only be promoted by pentasaccharide chains that contain *N*-acetyl glucosamine 6-*O*-sulfate → glucuronic acid → glucosamine *N*-sulfate 3-*O*-sulfate → iduronic acid 2-*O*-sulfate → glucosamine *N*-sulfate 6-*O*-sulfate (HajMohammadi et al. 2003) (Fig. 4.7). Clot-bound thrombin is resistant to inhibition by the heparin–antithrombin III complex and this may result in more thrombin generation in the circulation by the uninterrupted activation of factor V, VIII, and XI, thereby promoting extension of the intravascular clot leading to occlusive thrombus formation (Hong et al. 2006).

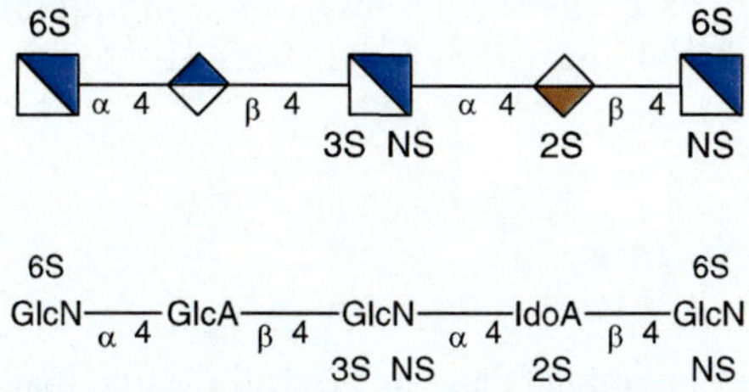

Fig. 4.7 Structure of a typical antithrombin binding pentasaccharide. IdoA, L-iduronic acid. GlcA, D-glucuronic acid. GlcN, glucosamine. 2S, 2-*O*-sulfate. 3S, 3-*O*-sulfate. 4S, 4-*O*-sulfate. 6S, 6-*O*-sulfate. NS, *N*-sulfated. This figure was created using GlycanBuilder (Ceroni et al. 2007)

The key reason behind the success of heparin in thrombosis and beyond is its polypharmacological sites of action for the prevention and treatment of multifactorial diseases that will only benefit slightly from agents based on a single pharmacological mechanism. Thromboembolic disorders are driven by hypercoagulable, hyperactive platelet, proinflammatory, endothelial dysfunction, and proangiogenesis states. Heparin can effectively modulate all of those multifactorial components, as well as the interface among those components (Harenberg 2009).

Low molecular weight heparin (LMWH) consists of fragments of heparin with a mean molecular weight of 5,000. Like heparin, it binds to antithrombin via a unique pentasaccharide sequence that is randomly dispersed in one-fifth of the LMWH chains. Less than half of the chains of LMWH have sufficient length to bridge antithrombin to thrombin. Consequently, LMWH preparations have greater activity against factor Xa than thrombin, with anti-Xa:anti-IIa ratios that range from 2:1 to 4:1 (Hirsh 1998).

LMWH has pharmacokinetic advantages over unfractionated heparin. LMWH produces a more predictable anticoagulant response and has a longer plasma half-life than unfractionated heparin. This permits fewer subcutaneous doses and obviates the need for laboratory monitoring. The incidence of HIT is lower with LMWH than with unfractionated heparin, which may offer a safety advantage (Rote et al. 1993). LMWH is more convenient to use than unfractionated heparin and this factor alone has led to the widespread use of LMWH in both arterial and venous thrombosis (Eikelboom et al. 2000).

Heparin-induced thrombocytopenia (HIT) is the most common disease associated with prolonged treatment with heparin. It is an immune-mediated prothrombotic disease that can lead to life-threatening thromboembolic complications. In the USA, it is estimated that 6 million new cases occur a year, with 3 million patients developing thrombotic complications and 90,000 deaths. HIT is divided into HIT I, a nonimmune-mediated drop in platelet count that occurs in heparin treated patients, and HIT II, the immune disease. HIT II is a disease mediated by antibodies to the complex of heparin and platelet factor 4 (PF4) (Kelton 2005). Heparin binding to PF4 induces a conformational change in PF4 exposing antigenic epitopes that generate antiheparin-PF4 antibodies (Horsewood et al. 1996). These antibodies bind to platelets and activate them. Activated platelets release prothrombotic particles and promote excessive thrombin generation and platelet consumption thrombocytopenia that is frequently followed by a prothrombotic state and antibody-mediated endothelial injury (Levy and Hursting 2007).

4.5.2 Heparan Sulfate

Heparan sulfate (HS) is a unique class of macromolecule that is widely expressed on the cell surface and in the ECM (see above). It is considered to be an analog of heparin. Cultured endothelial cells synthetize HS and 1–10% of these molecules have anticoagulant activity related to the presence of 3-*O*-sulfated glucosamine

(Weitz 2003). Heparan sulfate is related to several roles like blood coagulation, embryonic development, inflammatory responses, and viral and bacterial infections (Rosenberg et al. 1997; Esko and Selleck 2002; Liu and Thorp 2002; Taylor and Gallo 2006). HS is able to bind several fibroblast growth factors (FGFs) and their receptors, this binding is related to the correct embryonic development and cell proliferation (Wu et al. 2003).

Heparan sulfate is able to bind AT and this binding increases the ability of AT to bind and inhibit thrombin. Binding of heparan sulfate to AT leads to a conformational change in AT. This change leads to the generation of the active form of AT and inhibition of blood coagulation factors like factor Xa and thrombin (IIa). Factor Xa or thrombin binding to AT results in the cleavage of the P1–P1′ scissle bond in the reactive center loop (RCL) trapping the coagulant protease in a covalently bound inactive intermediate with the AT molecule (Gettins 2002) (Fig. 4.8).

Antithrombin has a binding site that interacts with a pentasaccharide sequence that contains a 3-*O*-sulfate glucosamine as the middle saccharide unit of the pentasaccharide. Surprisingly, mice deficient in 3-*O*-sulfotransferase 1, the enzyme that is responsible for 3-*O* sulfation of the middle residue of the pentasaccharide, display a normal phenotype, even though they have a reduced capacity to bind antithrombin and to enhance its reaction with factor Xa. These animals have normal fibrin deposition and normal thrombosis time (HajMohammadi et al. 2003). An explanation is that these knockout animals have a 2–20% residual activity of 3-*O*-sulfotransferase 1. This residual activity may be sufficient to protect these animals against a thrombotic phenotype. These animals also have activity of 3-*O*-sulfotransferase isoforms as 3-OST-6, so perhaps an animal that is completely null for 3-*O*-sulfotransferase may present a thrombotic phenotype (Weitz 2003).

4.5.3 Dermatan Sulfate

Dermatan sulfate (DS) is synthesized as a repeating polymer of D-glucuronic acid (GlcA) and *N*-acetyl-D-galactosamine (GalNAc), which is then modified by

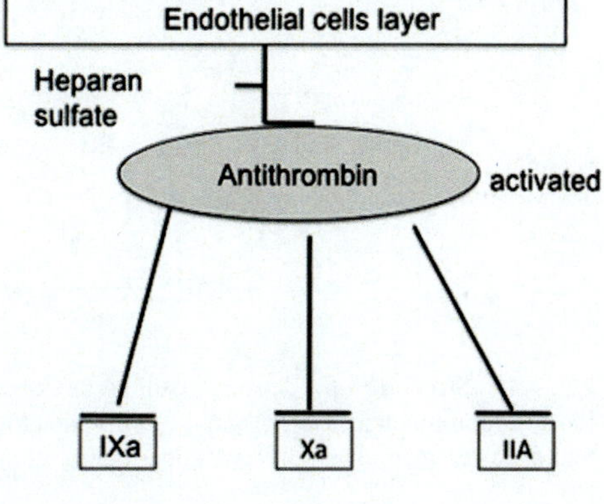

Fig. 4.8 Binding of circulating antithrombin to heparan sulfate localized in endothelial cells and in the medial layer of the vessel activates it, promoting inhibition of factors IXa, Xa, and IIa (thrombin)

epimerization of GlcA to L-iduronic acid (IdoA), sulfation of the 2-OH group of IdoA, and sulfation of the 4-OH and/or 6-OH groups of GalNAc (Trowbridge and Gallo 2002). Dermatan sulfate is the main activator of heparin cofactor II (HCII) and is unable to activate antithrombin. HCII by itself inhibits thrombin at a very slow rate, but the rate increases dramatically in the presence of heparin, heparan sulfate, or dermatan sulfate. For example, binding of DS to HCII increases inhibition of thrombin in 1,000× (Tollefsen et al. 1983) (Fig. 4.9). Human HCII is a single-chain glycoprotein containing 480 amino acid residues. It is homologous to antithrombin and other members of the serpin (serine protease inhibitors) family, with which it shares 30% amino acid sequence identity. The smallest fragment of porcine skin DS that binds to HCII with high affinity is a hexasaccharide containing IdoA 2-*O*-sulfate-3 → GalNAc 4-*O*-sulfate disaccharide subunits, which comprise only 5% of the total disaccharides present DS (Maimone and Tollefsen 1988) (Fig. 4.10). The ability of DS to bind and activate HCII is not simply a function of its overall charge density but appears to require the presence of GalNAc 4-*O*-sulfate

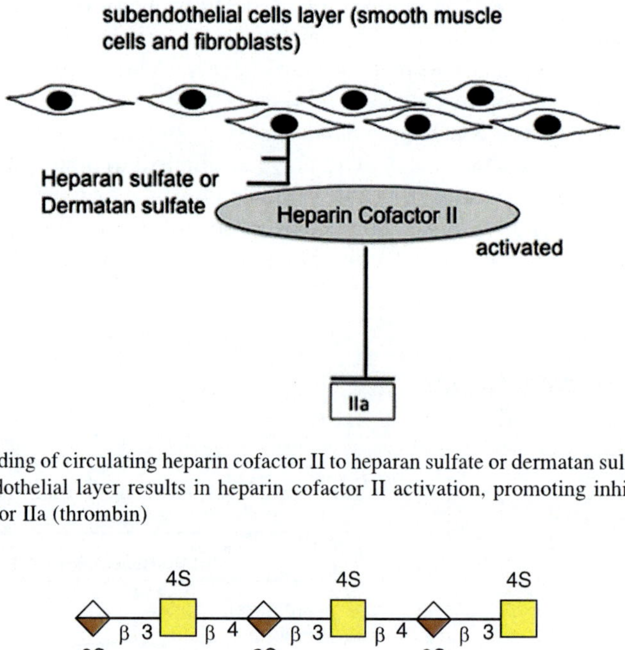

Fig. 4.9 Binding of circulating heparin cofactor II to heparan sulfate or dermatan sulfate localized in the subendothelial layer results in heparin cofactor II activation, promoting inhibition exclusively of factor IIa (thrombin)

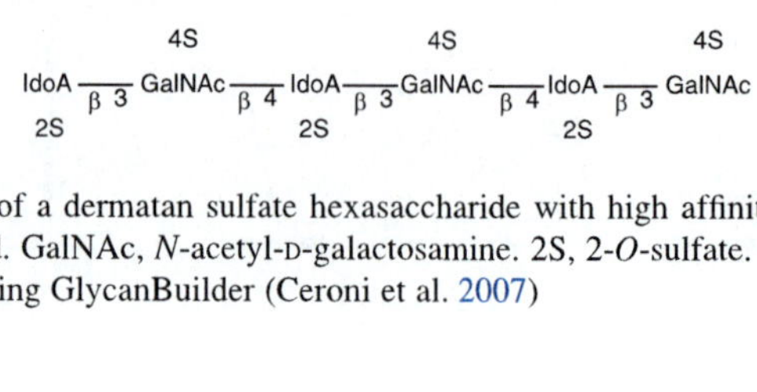

Fig. 4.10 Structure of a dermatan sulfate hexasaccharide with high affinity to heparan sulfate. IdoA, L-iduronic acid. GalNAc, *N*-acetyl-D-galactosamine. 2S, 2-*O*-sulfate. 4S, 4-*O*-sulfate. This figure was created using GlycanBuilder (Ceroni et al. 2007)

residues. It was observed that invertebrate DS polymers composed mainly of IdoA 2-*O*-sulfate-GalNAc 4-*O*-sulfate activate HCII at low concentrations, whereas polymers composed mainly of IdoA 2-*O*-sulfate-3 GalNAc6-*O*-sulafte are 1,000 times less active (Pavao et al. 1995, 1998). HCII activity is specific for thrombin, being unable to inhibit other proteins involved in coagulation or fibrinolysis. In contrast, antithrombin targets several coagulation proteases, including thrombin, factor Xa, and factor IXa (Tollefsen 2007). Heparan sulfate and dermatan sulfate are localized in different sites of the blood vessel. It was observed that DS is mainly localized in the adventitia but not in the media or intima layer and HS in mainly in the intima and media layers (Fig. 4.11). It was also observed that after endothelial injury HCII is observed in the adventitia colocalizing with DS. Most probably, this interaction is related with HCII activation by DS after arterial injury, since no HCII is localized in the adventitia before endothelial injury (He et al. 2008). HCII-deficient mice presented shorter thrombosis time when compared with wild-type animals. Dermatan sulfate injection was unable to prolong occlusion time in a thrombosis model mediated by photochemical-induced endothelial lesion. Injection of recombinant HCII in these animals was able to rescue normal thrombosis time in these animals and their sensitivity to DS. These results together demonstrated that the presence of HCII is essential for the antithrombotic effect of dermatan sulfate (He et al. 2002; Vicente et al. 2004).

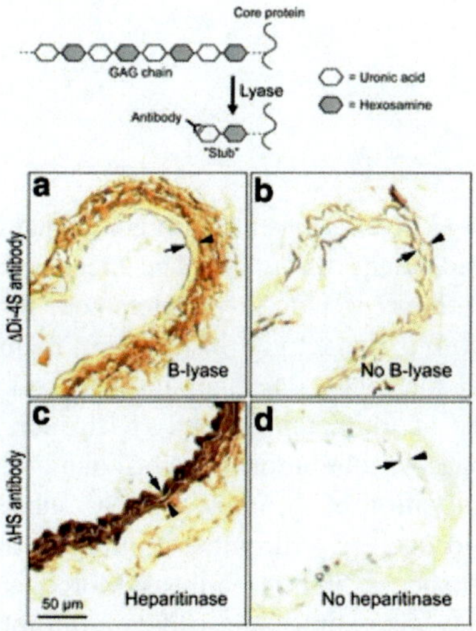

Fig. 4.11 Localization of dermatan sulfate and heparan sulfate in carotid arteries of wild-type mice. Frozen sections treated with chondroitin b-lyase (**a**) *Flaviobacterium* heparitinase (**c**) or buffer alone (**b**, **d**), and then incubated with monoclonal antibodies ΔDi-4S (**a**, **b**) or ΔHS (**c**, **d**). Bound antibodies were detected with peroxidase conjugated secondary antibody; arrows indicated external and internal elastic lamina. DS was present in adventitia and heparan sulfate in the intima/media. This research was originally published in He L, Giri TK, Vicente CP, Tollefsen DM (2008) Vascular dermatan sulfate regulates the antithrombotic activity of Heparin cofactor II. Blood 111 (8):4118–4125. Published with permission

4.5.4 Decorin

Decorin is expressed as an abundant small leucine-rich proteoglycan by a variety of cells, including cultured VSMCs, pericytes, and sprouting endothelial cells (Jarvelainen et al. 1991; Kaji et al. 2004). It was observed in the subendothelial matrix of healthy and diseased vessels (Merrilees et al. 2001). In contrast to the proteoglycans of normal blood vessels, the dermatan sulfate content of decorin is decreased during atherosclerosis (Shirk et al. 2000). This observation could explain the reduced ability of plaque-derived decorin to accelerate the inactivation of thrombin by heparin cofactor II. In vitro, the antithrombin activity of heparin cofactor II is potentiated by dermatan sulfate-enriched decorin or purified galactosaminoglycans, but not by chondroitin sulfate glycoforms or the isolated decorin core protein. Decorin has been shown to activate the thrombin inhibitor heparin cofactor II. Decorin inhibits fibrosis in animal models of response-to-injury-mimicking diseases such as intimal hyperplasia, pulmonary fibrosis, glomerulonephritis, corneal inflammation, and myocardial infarction (Isaka et al. 1996; Kolb et al. 2001; Nili et al. 2003; Schonherr et al. 2004; Weis et al. 2005). Decorin is associated with the regulation of collagen fibers assembly and organization in ECM and also may play a role in fibrin assembly and structure during thrombus formation. Decorin may also function as a bridge, stabilizing the thrombus against embolism (Dugan et al. 2006).

4.5.5 Intimatan

Intimatan (dermatan 4,6-*O*-disulfate) is a HCII agonist that inhibits thrombin in the fluid phase and is bound to the thrombus. The mechanism by which Intimatan prevents arterial rethrombosis is not fully understood. Earlier observations suggested that intimatan inhibits the activity of thrombin bound to fibrin, biomaterial surfaces (Hogg and Jackson 1989), and the site of vessel wall injury (Buchanan et al. 2001). The ability of intimatan to bind HCII upon thrombin binding to a thrombus suggested that the inhibition of clot-bound thrombin activity might prevent further local activation of thrombin and the subsequent accumulation of fibrin, thereby leading to occlusive thrombus formation, and that this effect may prevent reocclusion after successful thrombolysis. Moreover, the effect of intimatan on platelet reactivity may also contribute to its antithrombotic action through its ability to inhibit thrombin-induced platelet activation (Hennan et al. 2002). Although intimatan did not prevent ex vivo platelet reactivity in response to arachidonic acid or adenosine diphosphate, it did inhibit platelet aggregation by g-thrombin, a result not initially anticipated as g-thrombin lacks exosite 1 for HCII interaction (Rydel et al. 1994; Myles et al. 1998). The in vitro thrombin amidolytic assay verified that intimatan, but not dermatan sulfate, enhanced the inhibitory effect of HCII on g-thrombin. Although the mechanism is not yet elucidated, the higher charge density of intimatan specifically localized on the galactosamine ring may increase the affinity of the ternary GAG-HCII-g-thrombin complex allowing a more favorable interaction between the serpin and protease in the absence of a functional exosite usually required for HCII binding. Furthermore, intimatan was accompanied by a moderate increase in tongue bleeding time under conditions that ameliorate arterial rethrombosis after successful thrombolysis. The modest effect on bleeding time may be a reflection of intact platelet function mediated through the actions of arachidonic acid and adenosine diphosphate. Intimatan is able to prevent primary carotid arterial thrombosis, a protective effect that could not be achieved by the LMWH comparator drug, dalteparin. Intimatan is able to reduce the development of arterial rethrombosis after thrombolysis with a moderate increase of bleeding time and may be an effective antithrombotic agent for arterial thrombotic diseases such as coronary artery disease and ischemic stroke rather than just an anticoagulant for prevention of venous thrombosis (Hong et al. 2006).

4.6 Conclusions

GAGs are directly involved in pathological processes related to macrovascular disease, atherosclerosis, and thrombosis. The main role of GAGs in atherosclerosis is the accumulation of lipid retained by ECM molecules, especially by the chondroitin sulfate/dermatan sulfate proteoglycans versican, biglycan, and decorin. Retained lipid undergoes chemical modifications such as oxidation and oxidized LDL precipitates inflammatory responses such as monocyte recruitment and phagocytic uptake of LDL by differentiated monocytes (macrophages). These processes continue to eventually form a complex lesion with a necrotic core. The lesion may rupture and occlude the artery causing ischemic damage to organs and life-threatening clinical events such as heart attack and stroke. One therapeutic approach that we may consider to treat atherosclerosis is inhibiting the interaction of apoB with intimal proteoglycans. This involves direct blocking of the interaction between apoB and proteoglycans, and/or manipulating the synthesis of key retentive proteoglycans or their GAG side chains. For example, modifying proteoglycan synthesis and structure, such as GAG elongation and sulfation, may represent a target to prevent LDL binding and retention in the intima.

Hemostasis is a complex mechanism that controls blood fluidity, avoiding bleeding by the formation of a thrombus that stops the process. Thrombosis may be initiated by the intrinsic or extrinsic pathways of the coagulation cascade, both pathways converge in the activation of Factor X. Activated Factor X (Factor Xa) catalyzes formation of thrombin (Factor IIa) from prothrombin (Factor II). Inhibition of Xa and IIa thus play the most important role in triggering blood coagulation (Casu et al. 2004).

Blood maintains its fluidity mainly because of the action of natural inhibitors of the coagulation proteases and the antithrombotic effect of components of cells lining the vascular endothelium. The main natural inhibitors of blood coagulation are antithrombin and heparin cofactor II, two proteins belonging to the family of serpins, the predominant family of serine-protease inhibitors in humans. GAGs are responsible for the activation of AT and HCII, interaction of these factors with heparin, heparan sulfate, dermatan sulfate, decorin, and other biological or synthetic GAGs may increase inhibition of thrombin by these factors over 1,000 times. Also, alteration in the content of these molecules in endothelial or subendothelial layers of the vessels may alter their thrombotic characteristics, altering the thrombotic characteristic of the affected vessel.

There is sufficient evidence to warrant targeting proteoglycans of the vascular wall in the treatment of cardiovascular disease, thrombosis, and hemostasis. Recently, new anticoagulant and antithrombotic GAGs have been characterized from uncommon sources. A fucosylated chondroitin sulfate from sea cucumber presents antithrombotic and anticoagulant activity (Mourao et al. 1996; Pacheco et al. 2000). Also, heparin was obtained from test cells that has the same basic structure of mammalian heparin, but with less 2-*O* sulfation and is about 20 times less potent in the inhibition of thrombin by antithrombin when compared to the mammalian counterpart (Cavalcante et al. 2000). Dermatan sulfate from tissues of ascidians *S. plicata* and *A. nigra* (Chordate – Tunicate) (Pavao et al. 1995, 1998) were obtained and studied. *S. plicata* dermatan sulfate contains the same backbone with 2-*O*-sulfated α-L-iduronic acid and 4-*O*-sulfated *N*-acetyl-*β*-D-galactosamine residues and is a potent anticoagulant due to a high heparin cofactor II activity. In contrast, dermatan sulfate from *A. nigra* is also rich in 2-*O*-sulfated α-Liduronic acid, but in this case sulfated at *O*-6 of the *N*-acetyl-*β*-D-galactosamine units and has no in vitro or in vivo anticoagulant activity.

These discoveries may help in the development of new therapeutic agents to treat thrombosis and atherosclerosis in the early stages and preventing the dare consequences of these diseases in the next few years.

Acknowledgments This work was supported by Fundação de Amparo a Pesquisa do Estado de São Paulo – FAPESP 2007/01112-6, 2010/11474-5 (CPV) and 2009/00950-3, 2006/06560-4 (CCW), 2010/01119-3 (JAPG) and Conselho Nacional de Desenvolvimento Científico e Tecnológico-CNPq – 470608/2009-9 (CCW)

References

Aruffo A, Stamenkovic I et al (1990) CD44 is the principal cell surface receptor for hyaluronate. Cell 61(7):1303–1313

Ballinger ML, Nigro J et al (2004) Regulation of glycosaminoglycan structure and atherogenesis. Cell Mol Life Sci 61(11):1296–1306

Ballinger ML, Ivey ME et al (2009) Endothelin-1 activates ETA receptors on human vascular smooth muscle cells to yield proteoglycans with increased binding to LDL. Atherosclerosis 205(2):451–457

Bernfield M, Gotte M et al (1999) Functions of cell surface heparan sulfate proteoglycans. Annu Rev Biochem 68:729–777

Bobik A, Grooms A et al (1990) Growth factor activity of endothelin on vascular smooth muscle. Am J Physiol 258(3 Pt 1):C408–C415

Bowe MA, Mendis DB et al (2000) The small leucine-rich repeat proteoglycan biglycan binds to alpha-dystroglycan and is upregulated in dystrophic muscle. J Cell Biol 148(4):801–810

Brownlee M (2001) Biochemistry and molecular cell biology of diabetic complications. Nature 414(6865):813–820

Buchanan MR, Maclean GA et al (2001) Selective and sustained inhibition of surface-bound thrombin activity by intimatan/heparin cofactor II and its relevance to assessing systemic anticoagulation in vivo, ex vivo and in vitro. Thromb Haemost 86(3):909–913

Camejo G, Olsson U et al (2002) The extracellular matrix on atherogenesis and diabetes-associated vascular disease. Atheroscler Suppl 3(1):3–9

Cardoso LE, Mourao PA (1994) Glycosaminoglycan fractions from human arteries presenting diverse susceptibilities to atherosclerosis have different binding affinities to plasma LDL. Arterioscler Thromb 14(1):115–124

Casu B, Guerrini M et al (2004) Structural and conformational aspects of the anticoagulant and anti-thrombotic activity of heparin and dermatan sulfate. Curr Pharm Des 10(9):939–949

Cavalcante MC, Allodi S et al (2000) Occurrence of heparin in the invertebrate styela plicata (Tunicata) is restricted to cell layers facing the outside environment. An ancient role in defense? J Biol Chem 275(46):36189–36186

Ceroni A, Dell A et al (2007) The GlycanBuilder: a fast, intuitive and flexible software tool for building and displaying glycan structures. Source Code Biol Med 2:3

Chai S, Chai Q et al (2005) Overexpression of hyaluronan in the tunica media promotes the development of atherosclerosis. Circ Res 96(5):583–591

Chang MY, Potter-Perigo S et al (2000) Oxidized low density lipoproteins regulate synthesis of monkey aortic smooth muscle cell proteoglycans that have enhanced native low density lipoprotein binding properties. J Biol Chem 275(7):4766–4773

Chatzizisis YS, Coskun AU et al (2007) Role of endothelial shear stress in the natural history of coronary atherosclerosis and vascular remodeling: molecular, cellular, and vascular behavior. J Am Coll Cardiol 49(25):2379–2393

Clark RA (2001) Fibrin and wound healing. Ann N Y Acad Sci 936:355–367

Corsi A, Xu T et al (2002) Phenotypic effects of biglycan deficiency are linked to collagen fibril abnormalities, are synergized by decorin deficiency, and mimic Ehlers-Danlos-like changes in bone and other connective tissues. J Bone Miner Res 17(7):1180–1189

Cuff CA, Kothapalli D et al (2001) The adhesion receptor CD44 promotes atherosclerosis by mediating inflammatory cell recruitment and vascular cell activation. J Clin Invest 108(7):1031–1040

Cunningham KS, Gotlieb AI (2005) The role of shear stress in the pathogenesis of atherosclerosis. Lab Invest 85(1):9–23

Danielson KG, Baribault H et al (1997) Targeted disruption of decorin leads to abnormal collagen fibril morphology and skin fragility. J Cell Biol 136(3):729–743

Davignon J (2004) The cardioprotective effects of statins. Curr Atheroscler Rep 6(1):27–35

de Mattos DA, Stelling MP et al (2008) Heparan sulfates from arteries and veins differ in their antithrombin-mediated anticoagulant activity. J Thromb Haemost 6(11):1987–1990

Denis CV, Wagner DD (2007) Platelet adhesion receptors and their ligands in mouse models of thrombosis. Arterioscler Thromb Vasc Biol 27(4):728–739

Dietrich CP, Tersariol IL et al (1998) Structure of heparan sulfate: identification of variable and constant oligosaccharide domains in eight heparan sulfates of different origins. Cell Mol Biol (Noisy-le-grand) 44(3):417–429

Duan W, Paka L et al (2005) Distinct effects of glucose and glucosamine on vascular endothelial and smooth muscle cells: evidence for a protective role for glucosamine in atherosclerosis. Cardiovasc Diabetol 4:16

Dugan TA, Yang VW et al (2006) Decorin modulates fibrin assembly and structure. J Biol Chem 281(50):38208–38216

Edwards IJ, Xu H et al (1994) Interleukin-1 upregulates decorin production by arterial smooth muscle cells. Arterioscler Thromb 14(7):1032–1039
Edwards IJ, Wagner JD et al (2004) Arterial heparan sulfate is negatively associated with hyperglycemia and atherosclerosis in diabetic monkeys. Cardiovasc Diabetol 3:6
Eikelboom JW, Anand SS et al (2000) Unfractionated heparin and low-molecular-weight heparin in acute coronary syndrome without ST elevation: a meta-analysis. Lancet 355(9219):1936–1942
Endemann DH, Schiffrin EL (2004) Endothelial dysfunction. J Am Soc Nephrol 15(8):1983–1992
Endler G, Klimesch A et al (2002) Mean platelet volume is an independent risk factor for myocardial infarction but not for coronary artery disease. Br J Haematol 117(2):399–404
Esko JD, Lindahl U (2001) Molecular diversity of heparan sulfate. J Clin Invest 108(2):169–173
Esko JD, Selleck SB (2002) Order out of chaos: assembly of ligand binding sites in heparan sulfate. Annu Rev Biochem 71:435–471
Evanko SP, Raines EW et al (1998) Proteoglycan distribution in lesions of atherosclerosis depends on lesion severity, structural characteristics, and the proximity of platelet-derived growth factor and transforming growth factor-beta. Am J Pathol 152(2):533–546
Evanko SP, Angello JC et al (1999) Formation of hyaluronan- and versican-rich pericellular matrix is required for proliferation and migration of vascular smooth muscle cells. Arterioscler Thromb Vasc Biol 19(4):1004–1013
Fischer JW, Steitz SA et al (2004) Decorin promotes aortic smooth muscle cell calcification and colocalizes to calcified regions in human atherosclerotic lesions. Arterioscler Thromb Vasc Biol 24(12):2391–2396
Forbes JM, Yee LT et al (2004) Advanced glycation end product interventions reduce diabetes-accelerated atherosclerosis. Diabetes 53(7):1813–1823
Fraser JR, Laurent TC et al (1997) Hyaluronan: its nature, distribution, functions and turnover. J Intern Med 242(1):27–33
Furie B, Furie BC (2008) Mechanisms of thrombus formation. N Engl J Med 359(9):938–949
Gettins PG (2002) Serpin structure, mechanism, and function. Chem Rev 102(12):4751–4804
Gordillo GM, Sen CK (2003) Revisiting the essential role of oxygen in wound healing. Am J Surg 186(3):259–263
Gotlieb AI (2005) Atherosclerosis and acute coronary syndromes. Cardiovasc Pathol 14(4):181–184
Gozzo AJ, Nunes VA et al (2002) Glycosaminoglycans affect the action of human plasma kallikrein on kininogen hydrolysis and inflammation. Int Immunopharmacol 2(13–14):1861–1865
Greiling D, Clark RA (1997) Fibronectin provides a conduit for fibroblast transmigration from collagenous stroma into fibrin clot provisional matrix. J Cell Sci 110(Pt 7):861–870
Guretzki HJ, Gerbitz KD et al (1994) Atherogenic levels of low density lipoprotein alter the permeability and composition of the endothelial barrier. Atherosclerosis 107(1):15–24
HajMohammadi S, Enjyoji K et al (2003) Normal levels of anticoagulant heparan sulfate are not essential for normal hemostasis. J Clin Invest 111(7):989–999
Hallmann R, Horn N et al (2005) Expression and function of laminins in the embryonic and mature vasculature. Physiol Rev 85(3):979–1000
Harenberg J (2009) LMWH – new mechanisms of action. Thromb Res 123(Suppl 3):S1–S4
Hashimura K, Sudhir K et al (2005) Androgens stimulate human vascular smooth muscle cell proteoglycan biosynthesis and increase lipoprotein binding. Endocrinology 146(4):2085–2090
He L, Vicente CP et al (2002) Heparin cofactor II inhibits arterial thrombosis after endothelial injury. J Clin Invest 109(2):213–219
He L, Giri TK et al (2008) Vascular dermatan sulfate regulates the antithrombotic activity of heparin cofactor II. Blood 111(8):4118–4125
Hennan JK, Hong TT et al (2002) Intimatan prevents arterial and venous thrombosis in a canine model of deep vessel wall injury. J Pharmacol Exp Ther 301(3):1151–1156
Hirsh J (1998) Low-molecular-weight heparin for the treatment of venous thromboembolism. Am Heart J 135(6 Pt 3 Su):S336–S342
Hogg PJ, Jackson CM (1989) Fibrin monomer protects thrombin from inactivation by heparin-antithrombin III: implications for heparin efficacy. Proc Natl Acad Sci USA 86(10): 3619–3623

Hong TT, Van Gorp CL et al (2006) Intimatan (dermatan 4, 6-O-disulfate) prevents rethrombosis after successful thrombolysis in the canine model of deep vessel wall injury. Thromb Res 117 (3):333–342

Horsewood P, Warkentin TE et al (1996) The epitope specificity of heparin-induced thrombocytopenia. Br J Haematol 95(1):161–167

Huang F, Thompson JC et al (2008) Angiotensin II increases vascular proteoglycan content preceding and contributing to atherosclerosis development. J Lipid Res 49(3):521–530

Hultgardh-Nilsson A, Durbeej M (2007) Role of the extracellular matrix and its receptors in smooth muscle cell function: implications in vascular development and disease. Curr Opin Lipidol 18(5):540–545

Hurt-Camejo E, Olsson U et al (1997) Cellular consequences of the association of apoB lipoproteins with proteoglycans. Potential contribution to atherogenesis. Arterioscler Thromb Vasc Biol 17(6):1011–1017

Ignarro LJ, Buga GM et al (1987) Endothelium-derived relaxing factor produced and released from artery and vein is nitric oxide. Proc Natl Acad Sci U S A 84(24):9265–9269

Ilangumaran S, Briol A et al (1998) CD44 selectively associates with active Src family protein tyrosine kinases Lck and Fyn in glycosphingolipid-rich plasma membrane domains of human peripheral blood lymphocytes. Blood 91(10):3901–3908

Iozzo RV, Cohen IR et al (1994) The biology of perlecan: the multifaceted heparan sulphate proteoglycan of basement membranes and pericellular matrices. Biochem J 302(Pt 3):625–639

Isaka Y, Brees DK et al (1996) Gene therapy by skeletal muscle expression of decorin prevents fibrotic disease in rat kidney. Nat Med 2(4):418–423

Itano N, Sawai T et al (1999) Three isoforms of mammalian hyaluronan synthases have distinct enzymatic properties. J Biol Chem 274(35):25085–25092

Ivey ME, Little PJ (2008) Thrombin regulates vascular smooth muscle cell proteoglycan synthesis via PAR-1 and multiple downstream signalling pathways. Thromb Res 123(2):288–297

Jarvelainen HT, Kinsella MG et al (1991) Differential expression of small chondroitin/dermatan sulfate proteoglycans, PG-I/biglycan and PG-II/decorin, by vascular smooth muscle and endothelial cells in culture. J Biol Chem 266(34):23274–23281

Jennings LK (2009) Role of platelets in atherothrombosis. Am J Cardiol 103(3 Suppl):4A–10A

Jeziorska M, McCollum C et al (1998) Calcification in atherosclerotic plaque of human carotid arteries: associations with mast cells and macrophages. J Pathol 185(1):10–17

Kaji T, Sakurai S et al (2004) Characterization of chondroitin/dermatan sulfate proteoglycans synthesized by bovine retinal pericytes in culture. Biol Pharm Bull 27(11):1763–1768

Kallunki P, Tryggvason K (1992) Human basement membrane heparan sulfate proteoglycan core protein: a 467-kD protein containing multiple domains resembling elements of the low density lipoprotein receptor, laminin, neural cell adhesion molecules, and epidermal growth factor. J Cell Biol 116(2):559–571

Kelton JG (2002) Heparin-induced thrombocytopenia: an overview. Blood Rev 16(1):77–80

Kelton JG (2005) The pathophysiology of heparin-induced thrombocytopenia: biological basis for treatment. Chest 127(2 Suppl):9S–20S

Knudson W, Chow G et al (2002) CD44-mediated uptake and degradation of hyaluronan. Matrix Biol 21(1):15–23

Kolb M, Margetts PJ et al (2001) Proteoglycans decorin and biglycan differentially modulate TGF-beta-mediated fibrotic responses in the lung. Am J Physiol Lung Cell Mol Physiol 280(6): L1327–L1334

Kovanen PT, Pentikainen MO (1999) Decorin links low-density lipoproteins (LDL) to collagen: a novel mechanism for retention of LDL in the atherosclerotic plaque. Trends Cardiovasc Med 9(3–4):86–91

Kreuger J, Matsumoto T et al (2002) Role of heparan sulfate domain organization in endostatin inhibition of endothelial cell function. Embo J 21(23):6303–6311

Krumdieck R, Hook M et al (1992) The proteoglycan decorin binds C1q and inhibits the activity of the C1 complex. J Immunol 149(11):3695–3701

Kultti A, Rilla K et al (2006) Hyaluronan synthesis induces microvillus-like cell surface protrusions. J Biol Chem 281(23):15821–15828

Laurent TC, Fraser JR (1992) Hyaluronan. Faseb J 6(7):2397–2404

Levy JH, Hursting MJ (2007) Heparin-induced thrombocytopenia, a prothrombotic disease. Hematol Oncol Clin North Am 21(1):65–88

Libby P (2006) Inflammation and cardiovascular disease mechanisms. Am J Clin Nutr 83 (2):456S–460S

Lindahl U, Kusche-Gullberg M et al (1998) Regulated diversity of heparan sulfate. J Biol Chem 273(39):24979–24982

Little PJ, Tannock L et al (2002) Proteoglycans synthesized by arterial smooth muscle cells in the presence of transforming growth factor-beta1 exhibit increased binding to LDLs. Arterioscler Thromb Vasc Biol 22(1):55–60

Little PJ, Ballinger ML et al (2007) Vascular wall proteoglycan synthesis and structure as a target for the prevention of atherosclerosis. Vasc Health Risk Manag 3(1):117–124

Little PJ, Ballinger ML et al (2008) Biosynthesis of natural and hyperelongated chondroitin sulfate glycosaminoglycans: new insights into an elusive process. Open Biochem J 2:135–142

Liu J, Pedersen LC (2007) Anticoagulant heparan sulfate: structural specificity and biosynthesis. Appl Microbiol Biotechnol 74(2):263–272

Liu J, Thorp SC (2002) Cell surface heparan sulfate and its roles in assisting viral infections. Med Res Rev 22(1):1–25

Maimone MM, Tollefsen DM (1988) Activation of heparin cofactor II by heparin oligosaccharides. Biochem Biophys Res Commun 152(3):1056–1061

Malavaki C, Mizumoto S et al (2008) Recent advances in the structural study of functional chondroitin sulfate and dermatan sulfate in health and disease. Connect Tissue Res 49(3):133–139

Marcus AJ, Broekman MJ et al (2002) COX inhibitors and thromboregulation. N Engl J Med 347 (13):1025–1026

Matsuda M, Shimomura I et al (2002) Role of adiponectin in preventing vascular stenosis. The missing link of adipo-vascular axis. J Biol Chem 277(40):37487–37491

Medeiros GF, Mendes A et al (2000) Distribution of sulfated glycosaminoglycans in the animal kingdom: widespread occurrence of heparin-like compounds in invertebrates. Biochim Biophys Acta 1475(3):287–294

Merrilees MJ, Beaumont B et al (2001) Comparison of deposits of versican, biglycan and decorin in saphenous vein and internal thoracic, radial and coronary arteries: correlation to patency. Coron Artery Dis 12(1):7–16

Monzavi-Karbassi B, Stanley JS et al (2007) Chondroitin sulfate glycosaminoglycans as major P-selectin ligands on metastatic breast cancer cell lines. Int J Cancer 120(6):1179–1191

Mourao PA, Pereira MS et al (1996) Structure and anticoagulant activity of a fucosylated chondroitin sulfate from echinoderm. Sulfated fucose branches on the polysaccharide account for its high anticoagulant action. J Biol Chem 271(39):23973–23984

Murata K, Yokoyama Y (1982) Acidic glycosaminoglycan, lipid and water contents in human coronary arterial branches. Atherosclerosis 45(1):53–65

Myles T, Church FC et al (1998) Role of thrombin anion-binding exosite-I in the formation of thrombin-serpin complexes. J Biol Chem 273(47):31203–31208

Nader HB, Dietrich CP et al (1987) Heparin sequences in the heparan sulfate chains of an endothelial cell proteoglycan. Proc Natl Acad Sci U S A 84(11):3565–3569

Nader HB, Lopes CC et al (2004) Heparins and heparinoids: occurrence, structure and mechanism of antithrombotic and hemorrhagic activities. Curr Pharm Des 10(9):951–966

Nili N, Cheema AN et al (2003) Decorin inhibition of PDGF-stimulated vascular smooth muscle cell function: potential mechanism for inhibition of intimal hyperplasia after balloon angioplasty. Am J Pathol 163(3):869–878

Noonan DM, Fulle A et al (1991) The complete sequence of perlecan, a basement membrane heparan sulfate proteoglycan, reveals extensive similarity with laminin A chain, low density lipoprotein-receptor, and the neural cell adhesion molecule. J Biol Chem 266(34):22939–22947

Okamoto Y, Kihara S et al (2002) Adiponectin reduces atherosclerosis in apolipoprotein E-deficient mice. Circulation 106(22):2767–2770
Pacheco RG, Vicente CP et al (2000) Different antithrombotic mechanisms among glycosaminoglycans revealed with a new fucosylated chondroitin sulfate from an echinoderm. Blood Coagul Fibrinolysis 11(6):563–573
Palmer RM, Ferrige AG et al (1987) Nitric oxide release accounts for the biological activity of endothelium-derived relaxing factor. Nature 327(6122):524–526
Pavao MS, Mourao PA et al (1995) A unique dermatan sulfate-like glycosaminoglycan from ascidian. Its structure and the effect of its unusual sulfation pattern on anticoagulant activity. J Biol Chem 270(52):31027–31036
Pavao MS, Aiello KR et al (1998) Highly sulfated dermatan sulfates from Ascidians. Structure versus anticoagulant activity of these glycosaminoglycans. J Biol Chem 273(43):27848–27857
Philipson LH, Schwartz NB (1984) Subcellular localization of hyaluronate synthetase in oligodendroglioma cells. J Biol Chem 259(8):5017–5023
Pignone M, Phillips C et al (2000) Use of lipid lowering drugs for primary prevention of coronary heart disease: meta-analysis of randomised trials. BMJ 321(7267):983–986
Pillarisetti S, Paka L et al (1997) Endothelial cell heparanase modulation of lipoprotein lipase activity. Evidence that heparan sulfate oligosaccharide is an extracellular chaperone. J Biol Chem 272(25):15753–15759
Prehm P (1984) Hyaluronate is synthesized at plasma membranes. Biochem J 220(2):597–600
Rosenberg RD, Shworak NW et al (1997) Heparan sulfate proteoglycans of the cardiovascular system. Specific structures emerge but how is synthesis regulated? J Clin Invest 100(11 Suppl):S67–S75
Ross R (1999) Atherosclerosis is an inflammatory disease. Am Heart J 138(5 Pt 2):S419–S420
Rote WE, Werns SW et al (1993) Platelet GPIIb/IIIa receptor inhibition by SC-49992 prevents thrombosis and rethrombosis in the canine carotid artery. Cardiovasc Res 27(3):500–507
Rydel TJ, Yin M et al (1994) Crystallographic structure of human gamma-thrombin. J Biol Chem 269(35):22000–22006
Sakr SW, Potter-Perigo S et al (2008) Hyaluronan accumulation is elevated in cultures of low density lipoprotein receptor-deficient cells and is altered by manipulation of cell cholesterol content. J Biol Chem 283(52):36195–36204
Schmidt G, Hausser H et al (1991) Interaction of the small proteoglycan decorin with fibronectin. Involvement of the sequence NKISK of the core protein. Biochem J 280(Pt 2):411–414
Schonherr E, Sunderkotter C et al (2004) Decorin deficiency leads to impaired angiogenesis in injured mouse cornea. J Vasc Res 41(6):499–508
Segev A, Nili N et al (2004) The role of perlecan in arterial injury and angiogenesis. Cardiovasc Res 63(4):603–610
Seike M, Ikeda M et al (2006) Hyaluronan forms complexes with low density lipoprotein while also inducing foam cell infiltration in the dermis. J Dermatol Sci 41(3):197–204
Sen CK (2003) The general case for redox control of wound repair. Wound Repair Regen 11(6):431–438
Shirk RA, Parthasarathy N et al (2000) Altered dermatan sulfate structure and reduced heparin cofactor II-stimulating activity of biglycan and decorin from human atherosclerotic plaque. J Biol Chem 275(24):18085–18092
Skalen K, Gustafsson M et al (2002) Subendothelial retention of atherogenic lipoproteins in early atherosclerosis. Nature 417(6890):750–754
Srinivasan SR, Yost K et al (1980) Lipoprotein-hyaluronate associations in human aorta fibrous plaque lesions. Atherosclerosis 36(1):25–37
Stern R (2003) Devising a pathway for hyaluronan catabolism: are we there yet? Glycobiology 13(12):105R–115R
Stevens RL, Colombo M et al (1976) The glycosaminoglycans of the human artery and their changes in atherosclerosis. J Clin Invest 58(2):470–481
Stocker R, Keaney JF Jr (2004) Role of oxidative modifications in atherosclerosis. Physiol Rev 84(4):1381–1478

Stoll G, Bendszus M (2006) Inflammation and atherosclerosis: novel insights into plaque formation and destabilization. Stroke 37(7):1923–1932

Tabata T, Mine S et al (2007) Low molecular weight hyaluronan increases the Uptaking of oxidized LDL into monocytes. Endocr J 54(5):685–693

Taylor KR, Gallo RL (2006) Glycosaminoglycans and their proteoglycans: host-associated molecular patterns for initiation and modulation of inflammation. Faseb J 20(1):9–22

Thieszen SL, Rosenquist TH (1995) Expression of collagens and decorin during aortic arch artery development: implications for matrix pattern formation. Matrix Biol 14(7):573–582

Tollefsen DM (2007) Heparin cofactor II modulates the response to vascular injury. Arterioscler Thromb Vasc Biol 27(3):454–460

Tollefsen DM, Pestka CA et al (1983) Activation of heparin cofactor II by dermatan sulfate. J Biol Chem 258(11):6713–6716

Tovar AM, de Mattos DA et al (2005) Dermatan sulfate is the predominant antithrombotic glycosaminoglycan in vessel walls: implications for a possible physiological function of heparin cofactor II. Biochim Biophys Acta 1740(1):45–53

Tran PK, Tran-Lundmark K et al (2004) Increased intimal hyperplasia and smooth muscle cell proliferation in transgenic mice with heparan sulfate-deficient perlecan. Circ Res 94 (4):550–558

Tran PK, Agardh HE et al (2007) Reduced perlecan expression and accumulation in human carotid atherosclerotic lesions. Atherosclerosis 190(2):264–270

Tran-Lundmark K, Tran PK et al (2008) Heparan sulfate in perlecan promotes mouse atherosclerosis: roles in lipid permeability, lipid retention, and smooth muscle cell proliferation. Circ Res 103(1):43–52

Trowbridge JM, Gallo RL (2002) Dermatan sulfate: new functions from an old glycosaminoglycan. Glycobiology 12(9):117R–125R

Tumova S, Woods A et al (2000) Heparan sulfate proteoglycans on the cell surface: versatile coordinators of cellular functions. Int J Biochem Cell Biol 32(3):269–288

Vicente CP, He L et al (2004) Antithrombotic activity of dermatan sulfate in heparin cofactor II-deficient mice. Blood 104(13):3965–3970

Volpi N (2006) Therapeutic applications of glycosaminoglycans. Curr Med Chem 13(15):1799–1810

Wahab NA, Parker S et al (2000) The decorin high glucose response element and mechanism of its activation in human mesangial cells. J Am Soc Nephrol 11(9):1607–1619

Wakefield TW, Caprini J et al (2008) Thromboembolic diseases. Curr Probl Surg 45(12):844–899

Weigel PH, Hascall VC et al (1997) Hyaluronan synthases. J Biol Chem 272(22):13997–14000

Weis SM, Zimmerman SD et al (2005) A role for decorin in the remodeling of myocardial infarction. Matrix Biol 24(4):313–324

Weitz JI (2003) Heparan sulfate: antithrombotic or not? J Clin Invest 111(7):952–954

Whinna HC, Choi HU et al (1993) Interaction of heparin cofactor II with biglycan and decorin. J Biol Chem 268(6):3920–3924

Whitelock JM, Iozzo RV (2005) Heparan sulfate: a complex polymer charged with biological activity. Chem Rev 105(7):2745–2764

Wight TN (2002) Versican: a versatile extracellular matrix proteoglycan in cell biology. Curr Opin Cell Biol 14(5):617–623

Wight TN (2008) Arterial remodeling in vascular disease: a key role for hyaluronan and versican. Front Biosci 13:4933–4937

Wight TN, Merrilees MJ (2004) Proteoglycans in atherosclerosis and restenosis: key roles for versican. Circ Res 94(9):1158–1167

Wilkinson TS, Bressler SL et al (2006) Overexpression of hyaluronan synthases alters vascular smooth muscle cell phenotype and promotes monocyte adhesion. J Cell Physiol 206(2):378–385

Williams KJ, Tabas I (1995) The response-to-retention hypothesis of early atherogenesis. Arterioscler Thromb Vasc Biol 15(5):551–561

Wilson PG, Thompson JC et al (2008) Serum amyloid A, but not C-reactive protein, stimulates vascular proteoglycan synthesis in a pro-atherogenic manner. Am J Pathol 173(6):1902–1910

Winnemoller M, Schmidt G et al (1991) Influence of decorin on fibroblast adhesion to fibronectin. Eur J Cell Biol 54(1):10–17
Winnemoller M, Schon P et al (1992) Interactions between thrombospondin and the small proteoglycan decorin: interference with cell attachment. Eur J Cell Biol 59(1):47–55
Witztum JL (1994a) The oxidation hypothesis of atherosclerosis. Lancet 344(8925):793–795
Witztum JL (1994b) The role of oxidized LDL in the atherogenic process. J Atheroscler Thromb 1(2):71–75
Wu ZL, Zhang L et al (2003) The involvement of heparan sulfate (HS) in FGF1/HS/FGFR1 signaling complex. J Biol Chem 278(19):17121–17129
Xu T, Bianco P et al (1998) Targeted disruption of the biglycan gene leads to an osteoporosis-like phenotype in mice. Nat Genet 20(1):78–82
Yamaguchi Y, Mann DM et al (1990) Negative regulation of transforming growth factor-beta by the proteoglycan decorin. Nature 346(6281):281–284
Yao EH, Fukuda N et al (2008) Effects of the antioxidative beta-blocker celiprolol on endothelial progenitor cells in hypertensive rats. Am J Hypertens 21(9):1062–1068
Yoshino G, Hirano T et al (1997) Effect of long-term exogenous hyperinsulinemia and fructose or glucose supplementation on triglyceride turnover in rats. Atherosclerosis 129(1):33–39
Zhou Z, Wang J et al (2004) Impaired angiogenesis, delayed wound healing and retarded tumor growth in perlecan heparan sulfate-deficient mice. Cancer Res 64(14):4699–4702

Chapter 5
Glycans in Embryonic Stem Cells

Claire Pickford, Rebecca Holley, Kate Meade, and Catherine Merry

Abstract Heparan sulphate proteoglycans (HSPGs) are important modulators of cell fate during development via their roles in regulating growth factor signalling and morphogen gradients. The hypervariable sulphated structure of the heparan sulphate (HS) chain is chiefly responsible for the interaction with protein ligands. This structure is achieved through the action of a large cohort of modification enzymes. Mouse knockouts of many biosynthetic enzymes have been generated and illustrate the critical requirement for correctly patterned HS in early development and into adulthood. Embryonic stem (ES) cells are a malleable *in vitro* model for studying HS in early developmental events. In this chapter, we describe the essential role of HS in exit from pluripotency and early differentiation of mouse ES cells. Mouse ES cells produce a poorly sulphated HS that may protect from pro-differentiation cues (e.g. fibroblast growth factors). A HS epitope was found to mark cells with hemangioblast potential more precisely than the transcription factors or tyrosine kinase receptors previously used. Therefore, HS sequences can identify functionally unique populations of cells and so have potential applications in the development of cell-replacement therapies for degenerative conditions.

5.1 Glycosaminoglycans

Glycosaminoglycans (GAGs) are a species of complex polysaccharides. Members of the GAG family include chondroitin and dermatan sulphate (CS/DS), keratan sulphate (KS), hyaluronan (HA) and heparan sulphate (HS). With the exception of HA, most GAGs are covalently attached to a protein core to form a proteoglycan, and in general, the majority of the ligand-binding functions of proteoglycans are mediated by their GAG side chains (Gallagher 2001). HS has a particularly complex, hypervariable yet organised structure. HS proteoglycans (HSPGs) are located at the cell surface and in the extracellular matrix (ECM), and their strategic

C. Pickford, R. Holley, K. Meade, and C. Merry (✉)
School of Materials, Materials Science Centre, University of Manchester, Manchester, UK
e-mail: Catherine.merry@manchester.ac.uk

M.S.G. Pavão (ed.), *Glycans in Diseases and Therapeutics*, Biology of Extracellular Matrix,
DOI 10.1007/978-3-642-16833-8_5,

location in the pericellular domain indicates that they are key players at the interface between a cell and its microenvironment. This is reflected by the fact that HS chains interact with a large variety of proteins and cytokines (Gallagher 2001), with consequences in the biological function of the cell/tissue in health and disease. Knocking out the enzymes responsible for the initiation of HS biosynthesis, a complex process which will be discussed in more detail later, results in pre-gastrulation lethality in mice (Lin et al. 2000; Stickens et al. 2005). Consequently, the only cell or tissue type available for study from these mice are the embryonic stem (ES) cells, which are derived from pre-implantation blastocysts (Evans and Kaufman 1981). This chapter discusses HS biosynthesis and function, ES cell biology, why ES cells are an attractive model for understanding the former and some of the roles that HS has been discovered to play in ES biology and early differentiation events.

5.1.1 Heparan Sulphate Proteoglycans: Structure and Biosynthesis

HSPGs are found in all complex multicellular organisms, from *Caenorhabditis elegans* and *Drosophila melanogaster* through to higher mammals. There are three different classes of HSPGs, the transmembrane syndecans (four members), the GPI-anchored glypicans (six members), and the ECM molecules such as agrin, perlecan and collagen XVIII (Miosge et al. 2003; Noonan et al. 1991; Tsen et al. 1995). These may variously carry HS or CS/DS GAG chains. GAG chains are assembled sequentially onto proteins in the Golgi apparatus at consensus sequence sites. A tetrasaccharide linker is first added, then HS or CS/DS chain extension occurs. HS is a long linear polysaccharide whose structure consists of a backbone of alternating glucuronic acid (GlcA)-*N*-acetylglucosamine (GlcNAc) disaccharides $[\text{GlcA–GlcNAc}]_n$, which are then subjected to a series of modifications. These include deacetylation and N-sulphation of GlcNAc residues, conversion of GlcA to its epimer Iduronic acid (IdoA), 2-O-sulphation of GlcA/IdoA units, 6-O-sulphation and more rarely 3-O-sulphation of GlcNAc/GlcNS/N-unsubstituted glucosamine (GlcN) units (Esko and Lindahl 2001). All of these modifications are incomplete throughout the chain and are generated by a large bank of enzymes, many of which are members of multi-isoform families, which results in a wide diversity of structures, as illustrated in Fig. 5.1.

The N-sulphation of glucosamine is generally regarded as a prerequisite for subsequent modifications, which leads to clustering of modified regions and the generation of the typical domain structure of HS, based on the distribution of GlcNS-disaccharides, in which clustered regions of dense sulphation (S-domains) are interspersed along the backbone of unmodified N-acetylated regions (NA domains), flanked by regions of alternating N-sulphated and non-N-sulphated disaccharides (NA/NS-domains or transition zones) (Fig. 5.2) (Maccarana et al. 1996;

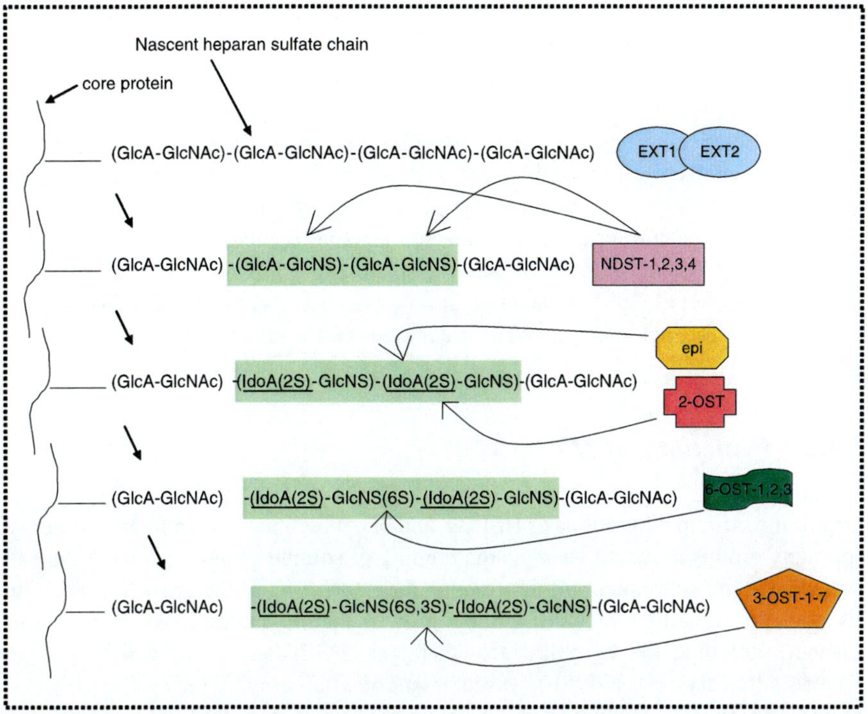

Fig. 5.1 Simplified cartoon of HS Biosynthesis. A complex of Ext1/2 enzymes catalyses the extension of the non-sulphated backbone. *N*-deacetylase/*N*-sulfotransferase (NDST) enzymes modify a proportion of GlcNAc to GlcNS. The C5 epimerase converts GlcA to its epimer IdoA. IdoA residues can be 2-sulphated by 2-O-sulfotransferase (2-OST), and GlcNS residues can be 6-O- and 3-O-sulphated by the 6-OST and 3-OST enzymes, respectively

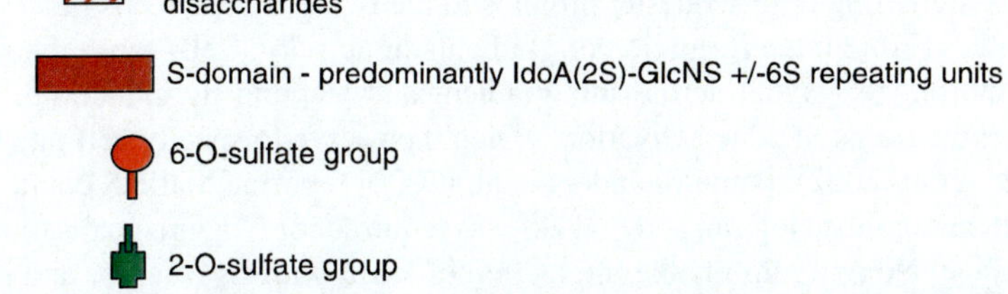

Fig. 5.2 Cartoon of HS domain organisation. HS is arranged into blocks of "composite sulphated domains" which comprise NS-domains flanked by transition zones, spaced between longer GlcA–GlcNAc containing sequences (N-acetylated domains). Adapted from Murphy et al. (2004)

Murphy et al. 2004). N-substitution patterns appear to be characteristic of the cell/tissue type of origin (Esko and Lindahl 2001; Ledin et al. 2004; Maccarana et al. 1996) rather than the organism in which that tissue is found.

More recently it has been discovered, in multiple species, that HS structure is edited post-biosynthetically at the cell surface by a family of endosulphatase enzymes (Sulfs) that preferentially de-6-O-sulphate HS (Dhoot et al. 2001). QSulf1 (from quail, the archetypal Sulf) is implicated in the positive regulation of Wnt signalling (Ai et al. 2003; Dhoot et al. 2001) and the negative regulation of FGF signalling (Wang et al. 2004). Expression of QSulf is spatiotemporally regulated across development (Dhoot et al. 2001; Zhao et al. 2006), as is 6-sulphation. The Sulf family of enzymes therefore act to fine-tune HS structure once it is presented at the cell surface.

5.1.2 Functions of HS

One of the principal functions of HSPGs, and one which has been a focus of interest for many groups in recent years, is the binding of soluble growth factors at the cell surface and the subsequent influence upon the interaction of the growth factor with its cognate signalling receptor. Many growth factors interact with HS in this manner, including the fibroblast growth factor (FGF) family, hepatocyte growth factor/scatter factor (HGF/SF), vascular endothelial growth factor (VEGF) and more. The most well-characterised interaction is that with FGF-2 (Gallagher 1998; Turnbull and Gallagher 1993). HS binds both FGF-2 and the FGF receptor (FGFR), forming a ternary complex that is necessary for signalling. Selectivity was demonstrated for distinct HS oligosaccharides that recognise different FGF family members (Guimond et al. 1993; Pye et al. 1998; Rapraeger et al. 1994; Turnbull et al. 1992). A crucial feature is the recognition of a very specific sequence of disaccharides in the HS chain (Lyon and Gallagher 1998) highlighting the importance of the coordinated function of the array of HS biosynthetic enzymes that have been described. Interestingly, this belief has been challenged recently with the development of libraries of semi-synthetic and biological HS oligosaccharides with defined structures that have been used to probe selectivity amongst the FGFs for HS sequences with little success (Kreuger et al. 2005).

An equally important function of HS is to localise factors to the cell surface to facilitate signalling or to sequester proteins in the ECM. Morphogens function over long ranges through the formation of gradients of activity. Cells sense the strength of the morphogen signal across this gradient and respond by induction of dose-dependent patterns of gene activation, which then serve to specify cell fate (Paine-Saunders et al. 2002). Paine-Saunders et al. (2002) reported that HS could bind to the BMP antagonist Noggin, and that HS was required for Noggin localisation at the cell surface. Noggin inhibits the interaction of BMP with its receptor and remains functional in this regard whilst bound to HS. This HS–Noggin interaction is subject to the action of Sulf enzymes. This highlighted a potential additional mechanism of the formation of the BMP gradient by interaction with a HS-regulated gradient of an antagonist, which would result in an inverse BMP activity gradient. Traditionally, gradients involve the ligand and its receptor acting in a feedback loop (Lecuit and Cohen 1998). The addition of a HS co-receptor to this system should make the system more robust (Lander 2005), especially since there is evidence for HSPG expression being regulated by those gradients that HSPGs influence. For example, the Drosophila glypican Dally has been shown to be expressed in feedback with the BMP homologue Decapentaplegic (Dpp) (Fujise et al. 2003).

5.1.3 HS Biosynthetic Enzyme Mouse Knockout Studies

5.1.3.1 Chain Polymerisation and N-Sulphation

The HS co-polymerase consists of a heterodimeric complex of EXT-1 and EXT-2 enzymes, which extend the nascent HS chain. The abolition of either gene is sufficient to disrupt HS synthesis entirely. The murine EXT-1 knockout phenotype was reported in 2000 and displayed an embryonic lethal phenotype between embryonic day (E) 6.5–8.5 due to a failure of gastrulation (Lin et al. 2000). ES cell lines derived from *Ext-1*$^{-/-}$ blastocysts were completely deficient in HS formation. *Ext-2*$^{-/-}$ embryos similarly perish at E6.0 without undergoing gastrulation, with the *Ext-2*$^{-/-}$ ES cells failing to produce HS (Stickens et al. 2005). These studies demonstrate that endogenously produced HS is vital for early developmental events, since maternal HS presumably present in the niche could not rescue these embryos. There is a human condition, Hereditary Multiple Exostoses (HME), characterised by loss of an allele of *Ext-1* or *Ext-2* that results in inappropriate growth of long bones and predisposition to certain types of bone cancer. Conditional knockouts of *Ext-1* in the mouse brain revealed defects concomitant with mutations in several HS-binding morphogens and severe perturbation of axon guidance pathways (Inatani and Yamaguchi 2003).

There are four *N*-deacetylase *N*-sulfotransferase (NDST) enzymes that act to convert GlcNAc to GlcNS. *Ndst-1* and *Ndst-2* have a wide tissue distribution in development and adults, whilst *Ndst-3* and *-4* are more restricted (Aikawa et al. 2001). *Ndst-2* mutant mice are viable, fertile and have a largely normal HS structure in all tissues, but fail to synthesise sulphated heparin in mast cells (Forsberg et al. 1999). Conversely *Ndst-1* mutant mice demonstrated structural alteration of HS in most basement membranes and died neonatally due to a failure to secrete surfactant into the lungs (Ringvall et al. 2000). A proportion of the embryos also exhibited cerebral hypoplasia and craniofacial defects consistent with impaired Sonic Hedgehog and FGF interaction with HS (Grobe et al. 2005). These data taken together indicated that NDST-1 was chiefly responsible for the sulphation of HS and NDST-2 for heparin, but that NDST-2 can, in a limited fashion, compensate for loss of NDST-1 in HS biosynthesis. These findings collectively emphasise the importance of correctly sulphated, spatiotemporally located HS in developmental patterning (Forsberg and Kjellen 2001).

Ndst-3 (Aikawa and Esko 1999) and *Ndst-4* (Aikawa et al. 2001) both display more restricted patterns of tissue expression than either *Ndst-1* or *-2*. *Ndst-3*$^{-/-}$ mice (Pallerla et al. 2008) are viable and fertile and show subtle abnormalities consistent with mild HS undersulphation. *Ndst2*$^{-/-}$/*Ndst3*$^{-/-}$ mice develop normally, showing that both isoforms are not essential for development. In contrast, *Ndst1*$^{-/-}$/*Ndst3*$^{-/-}$ embryos display developmental defects caused by severe HS undersulphation, demonstrating NDST-3 contribution to HS synthesis in the absence of NDST-1 (Pallerla et al. 2008).

The HS from ES cells generated from *ndst-1*$^{-/-}$/*ndst-2*$^{-/-}$ blastocysts is devoid of N-sulphation but retains a degree of 6-O-sulphation and contains a similar proportion of N-unsubstituted glucosamine (GlcN) units to wild-type HS. Since the prevailing view is that N-substitution generates substrates for the subsequent modification enzymes, this finding raises questions about the independence of the 6-O-sulphation modification step and also the origin of GlcN, which was thought to be generated by incomplete action of NDST isoforms (Holmborn et al. 2004).

5.1.3.2 Epimerisation and 2-O-Sulphation

Mice have also been generated with *2-ost*$^{-/-}$ (lacks 2-sulphation of HS chains) and *HSepi*$^{-/-}$ (lacks IdoA residues in HS chains) phenotypes. Both mutants exhibited bilateral renal agenesis, along with certain skeletal deformities, however the remainder of development was surprisingly unaffected (Bullock et al. 1998; Li et al. 2003). Analysis of HS from the embryonic fibroblasts of *2-ost*$^{-/-}$ showed a complete deficiency in 2-O-sulphation. However, the overall domain structure of HS was conserved, as well as the charge density maintained by increases in N- and 6-O-sulphation (Merry et al. 2001). Similar findings were reported for HS from *HSepi*$^{-/-}$ mice (Li et al. 2003). Interestingly, although mutant HS showed reduced binding affinity for FGF-1 and FGF-2 in a filter-binding assay, *2-ost*$^{-/-}$ fibroblasts could still mount a signalling response to these factors as assessed by ERK phosphorylation (Merry et al. 2001). This challenged the belief that IdoA(2S) residues were essential for FGF activation, and also demonstrated that HS does not have to bind growth factors with high affinity in order to promote signalling. The alteration in HS triggered by the loss of 2-OST demonstrates that the HS modifications are not independent of one another, but that aberrations in one step in the pathway can have broader effects elsewhere.

5.1.3.3 6-O-Sulphation

6-*O*-sulfotransferases (6OST) catalyse the addition of a sulphate group to C6 of GlcNAc/S sugars. Higher mammals have three isoforms, all of which show tissue specific expression patterns (Habuchi et al. 2000). In situ hybridisation studies have demonstrated that there is dynamic expression of the isoforms across development with differential expression in sites where HS-binding growth factors are known to be critical (Sedita et al. 2004). A larger in situ study that incorporated the *6-ost* isoforms along with *Ndst-1–4* and four of the *3-ost* isoforms (to be discussed below) also demonstrated spatio-temporal regulation but did reveal that many isoforms are expressed by the same cell/tissue at the same point in development, demonstrating that the enzyme complement of a cell may not be the sole determining factor in generating tissue-specific HS motifs (Yabe et al. 2005).

There have been two distinct studies described in relation to 6-OST-1 mutant animals, the genetrap mouse (Pratt et al. 2006) and the knockout mouse (Habuchi et al. 2007). The former displayed significant retinal axon guidance phenotypes (as did the *hs2st*$^{-/-}$ mouse, also a genetrap deletion). Habuchi and colleagues reported perinatal lethality in a proportion of *hs6st1*$^{-/-}$ animals, with surviving litters characterised as smaller than wild-type with placental vascular phenotypes. HS from these animals was found to have marked reduction in GlcNAc(6S) disaccharides. This was more severe in tissues known to have a higher expression of *6ost-1* over *6ost-2* and vice versa. Possible explanations for the relative mildness of the 6-OST-1 mutant phenotype are redundancy among isoforms in development, or possibly that regulation of 6-sulphate patterning within HS is largely the function of the Sulf enzymes discussed below.

5.1.3.4 6-O-Desulphation

mSulf1$^{-/-}$, *mSulf2*$^{-/-}$ and compound *mSulf1/2*$^{-/-}$ mice have been generated (Holst et al. 2007; Lamanna et al. 2006; Ratzka et al. 2008). Single and double knockout mice are viable, with *mSulf1*$^{-/-}$ and *mSulf2*$^{-/-}$ mice showing no overt phenotype. However, double knockout mice have a very short lifespan and an obvious reduction in bodyweight at birth. The contrast of the double knockout phenotype with single knockouts could reflect the overlapping expression pattern of mSulf1 and mSulf2 in the developing embryo, potentially allowing functional redundancy (Lum et al. 2007). However, the presence of one active *Sulf* allele is clearly essential for long-term viability of the organism.

5.1.3.5 3-O-Sulphation

The final, rarest modification of the HS chain is the addition of a sulphate group at the C-3 position of glucosamine, with seven members in mice identified to date (Shworak et al. 1999). Furthermore, multiple transcripts of the isoforms were found to be differentially expressed; 3-OST-1 (the original) was expressed in the kidney and brain, whereas 3OST-4 was restricted to brain, and 3-OST-3 exhibited a broader range of expression (Shworak et al. 1999). 3-OST-1 is implicated in the generation of the pentasaccharide that confers anticoagulant function to HS (Lindahl et al. 1980; Shworak et al. 1996). The *3-ost1*$^{-/-}$ mouse was expected to have a severe procoagulant phenotype because of this, but the mice were viable and fertile (HajMohammadi et al. 2003) and could maintain hemostasis (Shworak et al. 2002).

It seems that, even though *in vitro* studies did not detect anticoagulant activity in other *3-ost* isoforms, *in vivo* there is some functional redundancy and overlap (Girardin et al. 2005; Shworak et al. 2002).

Given that knocking out HS completely abolishes gastrulation, the restricted phenotypes of the enzyme knockouts described here was surprising, since IdoA (2S)-GlcNS disaccharides are thought to be essential for the ligand binding and activation properties of HS. For the *ndst-1*$^{-/-}$ and *HSepi*$^{-/-}$ embryos, it has been shown that structural changes in HS are common to all tissues (Ledin et al. 2004). There is clearly some redundancy in the NDST family, as many tissues in *ndst-1*$^{-/-}$ mice develop normally (Ringvall et al. 2000), although reported craniofacial abnormalities (Grobe et al. 2005) demonstrate that this is only partial. NDST-2 does not appear to contribute to the generation of HS structures in most tissues whilst NDST-1 is functional (Ledin et al. 2004) which is consistent with the restricted phenotype of the *ndst-2*$^{-/-}$ mice (Forsberg et al. 1999). The action of subsequent enzymes was thought to depend on substrate generation by NDSTs, yet HS from *ndst-1*$^{-/-}$/*ndst-2*$^{-/-}$ mice contains no N-sulphate groups but is still 6-sulphated (Holmborn et al. 2004). *Hs2st*$^{-/-}$ and *HSepi*$^{-/-}$ mice survive embryogenesis with most developmental pathways functioning appropriately (Bullock et al. 1998; Li et al. 2003). Most surprisingly the *3-ost1*$^{-/-}$ mice, expected to have severe perturbations of haemostasis, were viable (HajMohammadi et al. 2003; Shworak et al. 2002). RNAi technology now allows the possibility of simultaneous knockdown of all members of an isoform family, which would also be valuable in cases where one wished to abolish all 2O/6O/3O/N-sulphation or Sulf activity from a particular cell type *in vitro*.

5.1.3.6 Summary

It seems as though the early developmental events that critically depend on HS can proceed when the HS is mutated, presumably because the mutant HS has retained its ability to interact with growth factors and morphogens in a more or less wild-type fashion. However, the later embryonic or neonatal lethality of many of these mutations demonstrate that severe perturbations in HS structure are ultimately incompatible with life.

In terms of the regulation of HS structure, studies have demonstrated spatio-temporal expression of HS epitopes across development (David et al. 1992; van Kuppevelt et al. 1998), and also tissue-specific expression of the biosynthetic enzyme isoforms (Aikawa et al. 2001; Ford-Perriss et al. 2002), leading to the generally accepted hypothesis that the enzyme complement expressed in a cell/tissue will generate the tissue-specific structural features of HS. However, a study looking at the formation of FGF–HS–FGF–R ternary complexes in situ in whole embryo sections revealed that FGF–R binding to FGF–HS complexes altered dramatically over the course of one embryonic day across the whole embryo (Allen and Rapraeger 2003). This implies a global change in HS structure, which in turn suggests co-ordinate regulation across all tissues. At present, no mechanism has been suggested whereby the individual components of the HS biosynthetic machinery can be regulated this tightly and rapidly, therefore an open mind must be kept about the mechanism of regulation of HS structure, as it may involve hitherto uncharacterised factors.

5.2 Overview of ES Cell Biology

5.2.1 Deriving and Defining Embryonic Stem Cells

ES cells are derived following isolation and culture of the inner cell mass component of pre-implantation blastocysts (Evans and Kaufman 1981; Martin 1981; Thomson et al. 1998). The resulting ES cells are pluripotent and have the capacity to form all three germ layers of the developing embryo, as well as the extra-embryonic lineages of the trophectoderm and primitive endoderm (Fig. 5.3) (Niwa 2007).

ES cells are characterised by a rounded morphology and a high nucleus:cytoplasm ratio. Since ES cells undergo symmetrical division and therefore self-renew in culture, they have many useful applications to medical science, chiefly through the production of differentiated cells for tissue replacement therapy in degenerative disease, such as dopaminergic neurons to treat Parkinson's disease or β-islet cells to treat diabetes. However, they are also a valuable research tool in their own right, as they make understanding and manipulating the earliest developmental events

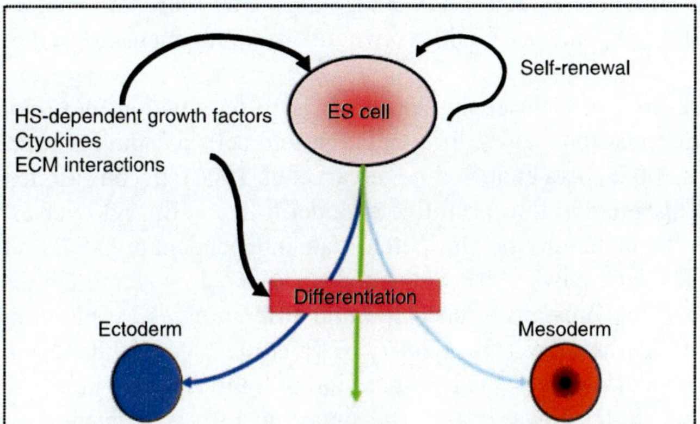

Fig. 5.3 Cartoon depicting mES cell pluripotency. ES cells will self-renew indefinitely given the right culture conditions e.g. mES cells require LIF. They retain the ability to differentiate into all three germ lineages under appropriate conditions

accessible *in vitro*. As discussed, HS has been implicated as having a crucial role in development of an organism (Lin et al. 2000; Stickens et al. 2005). HS has been shown to be a critical regulatory factor for many of the growth factors and morphogens that are involved in specific pathways during development (Smith et al. 2011). ES cells, with their ability to differentiate into multiple cell types, offer a valuable *in vitro* model system to investigate the roles of HS in defining cell identity and the progression of a cell population from an undifferentiated pluripotent state to a fully differentiated defined cell type. Furthermore, ES cell cultures can be readily expanded to provide sufficient HS for detailed structural analyses, and thus overcome one of the main obstacles of studying primary material.

Mouse ES cells were first isolated in 1981, human ES cells not until 1998, and whilst they are both pluripotent populations of cells and share many essential features, there are some important differences in their biology which shall be discussed below.

5.2.2 Pluripotency and Self Renewal

The pluripotent state of both mouse and human ES cells is sustained by a core regulatory network comprised primarily of transcription factors Oct4, Sox2 and Nanog. Each is expressed within the ICM and epiblast, and loss of any one component leads to a co-ordinated down regulation of all elements, leading to ES cell differentiation (Fong et al. 2008). Due to the inherent differences between mouse and human ES cells, the downstream targets of the network are not wholly comparable (Boyer et al. 2005; Loh et al. 2006). However, both human and mouse ES cells share a dependence on this core network for maintenance of the pluripotent state.

In mouse, loss of Oct4 expression leads to differentiation into trophectoderm, whilst over expression causes differentiation into cells resembling primitive endoderm (Niwa 2007; Niwa et al. 2000; Stewart et al. 2006). In contrast, loss of Nanog leads to differentiation into primitive endoderm-like cells and over expression in mouse ES cells maintains the pluripotent state independent of LIF/STAT3 pathway (Mitsui et al. 2003; Niwa 2007; Stewart et al. 2006). In a simplified model, it was proposed that Oct4 prevents trophectodermal differentiation by inhibiting transcription factors such as Cdx2 (Niwa 2007; Niwa et al. 2005), whilst Nanog prevents primitive endoderm expression by repression of transcription factors such as Gata6 (Mitsui et al. 2003; Niwa 2007). The opposing effects of Nanog and Oct4 on primitive endoderm differentiation can be linked to the autoregulation and feedback mechanisms involved in the transcriptional network. Oct4 partners with Sox2, forming a heterodimer that binds enhancers and modulates the expression of a number of genes including FGF4 and Nanog (Loh et al. 2006; Rodda et al. 2005; Stewart et al. 2006). The Oct4:Sox2 heterodimers and Nanog also regulate their own expression, creating a feedback loop which helps to maintain appropriate expression levels (Chew et al. 2005; Loh et al. 2006; Niwa 2007).

In both mouse and human ES cells, the three main transcription factors can behave as both activators and repressors of gene expression. As well as activating their own expression and that of genes involved in the maintenance of pluripotency, the transcription factors also act to repress genes implicated in early lineage specification (Boyer et al. 2005; Loh et al. 2006).

5.2.3 Culture of Mouse ES Cells

mES cells were first derived and maintained in serum-containing media on STO fibroblasts (Evans and Kaufman 1981; Martin 1981). It was discovered, however, that MEFs could be substituted for a single cytokine supplement, Leukaemia Inhibitory Factor, which was secreted by MEFs and is essential for maintenance of pluripotency *in vitro* (Smith et al. 1988; Williams et al. 1988). Likewise serum (FCS) can be substituted for knockout serum replacement (KSR, a defined synthetic substitute) (Ward et al. 2002). Recently, it has been found that an additional factor, bone morphogenic protein (BMP)-4, can in the presence of LIF eliminate the requirement for serum in culture regimens. BMP-4 results in Smad-phosphorylation and activation of Id genes which in turn prevent neural differentiation (Ying et al. 2003a). BMP4 also functions by blocking ERK phosphorylation and signalling via the MAPK pathway, events upstream of mouse ES cell differentiation (Qi et al. 2004; Ying et al. 2008). The refinement of the signalling mechanisms involved in maintaining mouse ES cell pluripotency has led to the proposed "ground state" model of ES cell self-renewal (Ying et al. 2008). By combining low concentrations of inhibitors of FGF receptor (FGFR) and ERK signalling with defined media, mouse ES cell pluripotency was maintained but with reduced cell viability. However, viability was restored by the addition of a GSK3 inhibitor which was hypothesised to compensate for the loss of the pro-proliferative signalling of the MAPK/ERK pathway (Ying et al. 2008). This led to the production of "3i" media that sustained ES cells independently of the LIF and BMP4. Therefore, Ying et al. proposed that ES cells inherently self-maintain their pluripotent phenotype unless stimulated by exogenous factors that result in down regulation of the core transcriptional network and subsequent differentiation. In comparison, signalling via LIF/STAT3 and BMP4 buffer the ES cells against exogenous stimulation in undefined culture conditions (i.e. endogenous FGF4 signalling via ERK) (Ying et al. 2008).

5.2.4 Culture of Human ES Cells

In contrast to mouse ES cells, human ES cells cannot be maintained by LIF and suppression of BMP signalling promotes human ES cell pluripotency (Xu et al. 2005a, b). Human ES cells were originally derived on a feeder layer of mouse embryonic fibroblasts (MEFs), providing a surface for attachment and secreting an unknown combination of factors into the surrounding media (Thomson et al. 1998). If human ES cells are to be applied therapeutically, it is necessary for a defined, xeno-free environment (free from feeder layers and animal derived products) to be determined, as highlighted by Martin et al. (2005) who demonstrated the incorporation of non-human sialic acid Neu5Gc by human ES cells cultured on a MEF feeder layer.

5.2.4.1 A Suitable Substrate

As well as secreting supporting factors into the media, MEFs provide a matrix of proteins and glycans, including fibronectin, collagen I, collagen IV and laminin, for human ES cell attachment, proliferation and self-renewal (Braam et al. 2008). The MEF feeder layer was initially replaced by Matrigel™, a heterogeneous mix of ECM molecules including collagen IV, HS proteoglycans (HSPGs) and laminin (Kleinman et al. 1982; Xu et al. 2001). Matrigel™ has subsequently supported the successful culture of human ES cells in a variety of different media compositions (Levenstein et al. 2006; Li et al. 2005; Lu et al. 2006; Ludwig et al. 2006a, b; Xu et al. 2005a, b). However, Matrigel™ is produced by mouse Engelbreth–Holm–Swarm (EHS) tumours and represents an undefined, animal-derived product in human ES cell culture.

In order to refine the culture substrate, studies have maintained human ES cells in a variety of different media on TCP coated with ECM substrates such as collagen IV, fibronectin and laminin. A number of studies have successfully used laminin as a substrate with the integrin dimer $\alpha 6\beta 1$ specific for laminin binding detected on the cell surface (Xu et al. 2001; Li et al. 2005). Although Xu et al. (2001) found fibronectin inferior to laminin, a number of studies have successfully maintained human ES cells on fibronectin (Amit et al. 2004) and in some cases have found laminin to be less successful than fibronectin (Baxter et al. 2009). The ability of fibronectin to support human ES cell growth is re-enforced by the expression and interaction of fibronectin-binding integrins such as $\alpha 5\beta 1$ (Baxter et al. 2009). In comparison, Ludwig et al. (2006a, b) determined that a combination of collagen IV, fibronectin, laminin and vitronectin was necessary to support cells in fully defined media (TeSR1).

5.2.4.2 Culture Media

The first step towards a more defined environment was the replacement of serum with a synthetic substitute, "Knockout Serum Replacement" (Amit et al. 2000; Ward et al. 2002). Developed for mouse ES cell culture, serum replacement is more defined, largely free of animal-derived products and does not suffer from the batch-to-batch variations associated with FCS. However, despite dramatically enhancing cloning efficiency, this substitution required the addition of relatively low concentrations (4 ng/ml) of FGF-2 to maintain proliferation of the clonally derived human

ES cells (Amit et al. 2000). From this study onwards, FGF-2 has been a key factor in the development of a defined culture environment and is present in all subsequent media compositions. The requirement for MEF feeder cells was partially bypassed by successfully culturing hES cells on human fibronectin in MEF conditioned media supplemented with FGF-2 (8 ng/ml) (Xu et al. 2001). However, higher concentrations of FGF-2 (100 ng/ml) are required to support human ES cell self-renewal and proliferation in unconditioned, KSR supplemented media (Levenstein et al. 2006; Xu et al. 2005a, b). Levenstein et al. (2008) proposed that the dramatic increase in FGF concentration necessary in unconditioned media may be partially attributed to the lack of stabilising HSPGs normally secreted by MEFs. The study identified five species of MEF-secreted HSPGs, including perlecan, agrin and syndecan-4, and proposed that the secreted products not only stabilised FGF-2 in the media but also assisted receptor binding (Levenstein et al. 2008). However, although addition of heparin/HS enhanced human ES cell culture at lower FGF-2 concentrations they did not match growth in MEF conditioned media, suggesting optimal FGF-2 signalling required a specific repertoire of MEF secreted HSPGs (Levenstein et al. 2008). In contrast, Furue et al. (2008) successfully applied heparin to a defined human ES cell culture system, maintaining human ES pluripotency at levels equivalent to growth on MEF feeder layers. The study highlighted the potential of heparin/HS in human ES cell propagation, reporting a dose dependent increase in colony formation and FGF receptor (FGFR) phosphorylation with added heparin. The optimal media composition contained 100 ng/ml heparin and 10 ng/ml FGF-2, a dramatically reduced concentration compared to previous studies. The differing results could be partially attributed to different media compositions. Whilst Levenstein et al. used DMEM/F12 media, Furue et al. applied a base medium, ESF, previously developed for mouse ES cell culture. A recent study indicated that although human ES cells express all four FGF receptors (FGFR, with FGFR1 expression predominant), FGF-2 signals primarily through FGFR2 (Eiselleova et al. 2009). The study also highlighted the expression of endogenous FGF-2 by human ES cells and proposed that FGF-2 functions by primarily promoting cell adhesion and survival rather than directly maintaining pluripotency. This is supported by Furue et al., who reported enhanced FGFR phosphorylation with added heparin in the absence of exogenous FGF-2 (Furue et al. 2008).

The concentration of FGF-2 in unconditioned media can be reduced when combined with other factors including TGFβ1 (Amit et al. 2004; Ludwig et al. 2006a, b), and the BMP inhibitor noggin supported human ES cells in media supplemented with KSR, 4 ng/ml FGF-2 and 0.12 ng TGFβ1 (Amit et al. 2004; Xu et al. 2005a, b). The two factors were also fundamental elements of the defined media TeSR1, which also substituted KSR with a supplement of lipids, minerals and cloned growth factors creating a completely defined, animal free culture medium in which two new human ES cell lines were derived (Ludwig et al. 2006a, b). In a study that brings to mind the hypothesised "ground state" of mouse ES cells, Xu et al. highlighted the need to suppress pro-differentiation signals such as BMP4 to sustain ES cell pluripotency (Xu et al. 2005a, b). Addition of noggin to unconditioned media reduced the required concentration of FGF-2 to

40 ng/ml. The study also revealed that MEF-conditioned media promotes a lower level of BMP signalling compared to unconditioned media, suggesting MEFs act to either remove the signalling agent or to secrete inhibitors into the media.

Despite progressive steps towards a defined culture environment, human ES cells are still routinely cultured with MEF-conditioned media or unconditioned media supplemented with high concentrations of FGF-2. Indeed, in many laboratories, precious stock cultures of human ES cells are maintained on MEFs with KSR and low FGF-2 levels. The inability to depart from reliance of MEFs is in part due to a lack of understanding of the molecular mechanisms maintaining human ES cells and also an inability to fully replicate the results presented in the above papers (Baxter et al. 2009).

5.3 The Role of HSPGs in mES Cell Early Lineage Events

Since one of the key goals of ES cell research is to generate defined cell types for tissue replacement therapy and the study of development, many protocols now exist for the directed differentiation of mES cells, enriching for one specific cell type or lineage. We have characterised the HSPG phenotype in mES cell neuroectodermal and mesodermal *in vitro* differentiation and discuss these data in light of current knowledge of ES cell biology (Fig. 5.3).

5.3.1 The Role of HS in Neuroectodermal Differentiation of mES Cells

Established protocols exist for the conversion of ES cells into neuroectodermal precursors in adherent monoculture. Neither aggregation nor co-culture is required for this, merely the elimination of inductive signals for alternative fates (Ying et al. 2003a, b). An ES cell line was engineered with a green fluorescent protein (GFP) gene knocked into the *sox1* locus (Sox1-GFP) (Aubert et al. 2003). Sox1 is the earliest known specific marker of neuroectodermal precursor cells (Wood and Episkopou 1999) and has a highly restricted pattern of expression – it is expressed in the neural plate and neuroepithelial cells but down-regulated during neuronal and glial differentiation. When cells are incubated in a serum-free growth medium optimised for neural cells, N2B27 medium (Ying et al. 2003a, b), up to 60% neural precursor conversion can be achieved by day 4. The GFP insert allows these precursors to be isolated from the undifferentiated cells by flow cytometry, and re-plated, after which they can differentiate efficiently into a range of neuronal cell types. In the absence of other inductive signals, endogenous FGF-4 is likely to be driving this differentiation, and it has been demonstrated that addition of exogenous FGF-4, under limiting autocrine conditions, will increase the conversion frequency (Ying et al. 2003a, b). Research into the very earliest signalling networks in this system have shown the importance of FGF-mediated Erk phosphorylation events (Stavridis et al. 2010; Stavridis et al. 2007). FGF-4 has been shown to have a requirement for HS to allow ternary complex formation (Guimond and Turnbull 1999; Rapraeger et al. 1994) which suggests an influential role for HS in this process.

Our initial study of HS in mES cells detailed a comparison of HS structure between pluripotent ES cells and Sox1+ neural precursor cells. We correlated the structural data with a large scale expression analysis of all the biosynthetic enzyme isoforms associated with HS assembly. Characterisation of HS purified from metabolically radiolabeled cultures revealed that mES cells produce HS chains with a typical domain structure, but these are poorly sulphated when compared to HS from other embryonic tissues (Ledin et al. 2004; Holmborn et al. 2004). The chains consisted of only 30–33% GlcNS-containing disaccharides. This increased significantly in Sox1+ cells, particularly manifest in an extension of contiguous regions of N-sulphation or S-domains. There were also increases in 6-sulphation and, to a lesser extent, 2-sulphation (Fig. 5.4).

In addition to changes in structure, we found quantitative increases in the expression levels of a variety of the sulfotransferases, with the most marked increases displayed by more restricted isoforms such as *Ndst-4*, *3-ost-3A* and *3-ost-5*, which had been virtually absent in ES cells (Johnson et al. 2007).

Our study also made use of a panel of sequence-specific anti-HS antibodies. These were generated by phage display and each recognises a unique three-dimensional structural motif within HS chains (van Kuppevelt et al. 1998). They have been previously used in immunohistochemical studies to visualise the distinct

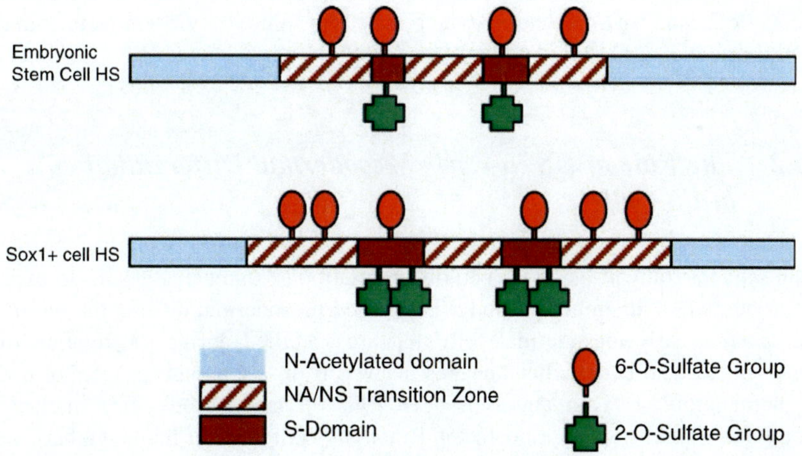

Fig. 5.4 Cartoon comparing the possible structure of ES cell HS with HS secreted from cells after 7 days of culture in N2B27 neural differentiation medium. N-sulphation increases within the composite regions of modification, namely the S-domains and the transition zones. Separately, 6-sulphation and 2-sulphation content increase, but there is no increase of the UA(2S)–GlcNS(6S) disaccharide

location of specific HS structures within tissues (ten Dam et al. 2003). In our work, we utilised this panel in non-destructive flow cytometric analyses to compare ES cells with Sox1+ cells, generating a cell-type specific "fingerprint" of HS epitope expression. Distinct epitopes were found to be up-regulated as a function of neural differentiation, reflecting the shift in HS patterning that was characterised biochemically. The advantage of flow cytometric analysis will become apparent in a further study described below.

Importantly, we also studied the *Ext-1*$^{-/-}$ HS-deficient ES cell line. These cells self-renew in standard ES culture conditions with no requirement for endogenously produced HS. However, they have an inability to differentiate effectively to Sox1+ neural precursors in N2B27 media (Johnson et al. 2007; Kraushaar et al. 2010; Pickford et al. 2011). The HS-deficient cells instead retain their markers of pluripotency. Addition of soluble heparin to the culture media allowed a partial rescue of neural differentiation. It should be noted that our observations concerning the *Ext1*$^{-/-}$ cell line have been challenged (Sasaki et al. 2008), although they are in good agreement with the original publications detailing their derivation (Lin et al. 2000; Stickens et al. 2005).

A recent study by Lanner et al. (2009) provided further evidence for the essential role of HS sulphation in ES cell differentiation. Mouse ES cells deficient in NDST1 and 2 and cells treated with $NaClO_3^-$ (a specific inhibitor of PAPS synthetase) also demonstrated impaired differentiation (Lanner et al. 2009). Cells treated with $NaClO_3^-$ demonstrated impaired FGF2 and FGF4 signalling and had enhanced Nanog expression compared to control cultures. Intriguingly, NDST1$^{-/-}$/2$^{-/-}$ cells demonstrated attenuated FGF4, but not FGF2 signalling, indicating a specific requirement for N-sulphation in FGF4 binding *in vivo*. However, the removal of NDST activity does not only remove N-sulphation but also has significant effects on 2- and 6-O-sulphation, which may also influence growth factor binding (Holmborn et al. 2004).

5.3.2 The Role of HS in Early Mesodermal Differentiation of ES Cells

A subsequent study utilising ES cell differentiation to study changes in HS expression focussed on an embryoid-body (EB) based mesodermal differentiation protocol. *In vivo*, early mesodermal cells that are specified during gastrulation form hematopoetic and endothelial lineages derived from a common progenitor called the hemangioblast (Tam and Behringer 1997). Early mesoderm is marked by expression of the transcription factor Brachyury (Bry), with hemangioblast cells additionally expressing Flk1, the VEGF-2 receptor. Flk1 expression is essential for establishment of further differentiated blood lineages. This differentiation sequence can be recapitulated via EBs *in vitro*. First, ES cells are re-plated in suspension in defined media where they proliferate to form three-dimensional balls of cells, the EB. A transient population of progenitors exist within the EB for approximately

36 h between day 2.5 and 4 of EB differentiation (Choi et al. 1998). These cells express both Bry and Flk1, and represent the *in vitro* equivalent of the yolk sac haemangioblast (Fehling et al. 2003). In response to VEGF, these cells can generate primitive and definitive hematopoietic, endothelial and vascular smooth muscle progeny (Kennedy et al. 1997). As with Sox1-GFP in the neural system, we utilised a reporter cell line with GFP under the control of the Bry promoter and cells were further phenotyped by flow cytometry using an antibody against Flk1. Flk1+ cells were found to have altered HS sulphation patterning from ES cells as demonstrated using the anti-HS monoclonal antibody 10E4 (Baldwin et al. 2008), which recognises a general sulphated motif (David et al. 1992), similar to that seen in neural differentiation (Johnson et al. 2007). Subsequently, ES cells were compared with hemangioblast cells using the HS–epitope specific panel of phage display antibodies. A variety of changes were observed, including a substantial increase in the binding of RB4EA-12, which requires 6-sulphation. Most significantly, antibody HS4C3 separated Bry+/Flk1+ hemangioblast cells into HS4C3 high ($HS4C3^{hi}$) and low ($HS4C3^{lo}$) binding populations (Baldwin et al. 2008). The specificity of HS4C3 is interesting: ability to bind relatively weakly to NS- and 6S-containing motifs, with high affinity binding triggered by an additional 3S group.

When standard VEGF-dependent EB re-plating assays were conducted on the separated populations, $HS4C3^{hi}$ cells were found to have a dramatically increased potential to form both blast and endothelial colonies compared to $HS4C3^{lo}$ cells. A time-course analysis of the appearance of $HS4C3^{hi}$ cells demonstrated that it was a transient phenotype and that cells become $Bry^{+}/HS4C3^{hi}$ before becoming $Bry^{+}/Flk1^{+}/HS4C3^{hi}$, defining the hemangioblast. Therefore, HS4C3 marks cells with maximal hemangioblast potential. The VEGF dependency of the colony-forming assays suggests that cells expressing the $HS4C3^{hi}$ epitope may have increased responsiveness to VEGF. Whilst VEGF is not an obligate HS-dependent growth factor, its interaction with Flk1 is strongly enhanced by the presence of HS (Terman et al. 1994; Tessler et al. 1994). The HS epitope required for VEGF/Flk1 binding and signalling remains unknown, although long HS chains containing a minimum of 22 saccharides are required *in vitro* (Robinson et al. 2006; Soker et al. 1994). The sequences identified in these studies are likely to contain 3-O-sulphate groups, however limitations in the biochemical techniques for 3-O-sulphated-HS analysis mean this is currently unknown. Taken together with our knowledge of the binding requirements of HS4C3, however, we hypothesise that HS4C3 binding with high affinity to cells marks HS chains with increased ability to promote VEGF signalling (Fig. 5.5). We believe this is the first demonstration of a sulphated motif within HS identifying a functionally unique population of cells. Importantly, we were able to correlate this finding with data from the early mouse embryo, with HS4C3-recognising HS restricted to early mesodermal tissues, overlapping with Flk1 expression.

Together, these studies suggest that the low sulphated HS expressed by mouse ES cells may limit the signalling of pro-differentiation factors, a theory that supports the "ground state" model of mouse ES cell pluripotency proposed by Ying et al. (2008). This hypothesis is also supported by the phenotype of $EXT1^{-/-}$ mouse ES cells. Despite lacking endogenous HS, $EXT1^{-/-}$ ES cells maintain a

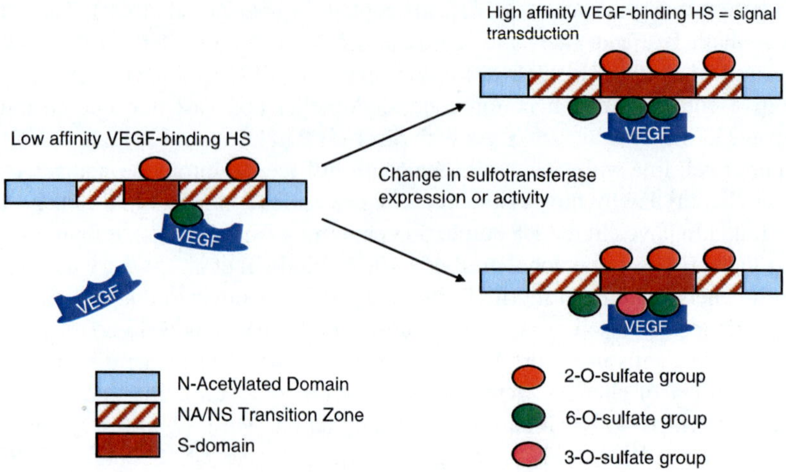

Fig. 5.5 Model depicting the hypothesised change in VEGF-binding potential of HS produced by hemangioblast cells identified through high affinity binding of HS4C3

pluripotent state (expressing Nanog and Oct4) and have characteristic ES cell morphology in culture. However, $EXT1^{-/-}$ cells undergo impaired EB formation (Holley et al. 2010), do not differentiate in the absence of LIF (Kraushaar et al. 2010; Lin et al. 2000) and cannot undergo neural differentiation in conditions proposed by Ying et al. (Johnson et al. 2007). It has been recently reported that this phenotype can be replicated by culturing wild-type cells in the presence of a specific inhibitor of FGFRs and that $EXT1^{-/-}$ cells have a reduced response to exogenous FGF2 indicating defective signalling via this pathway (Kraushaar et al. 2010). However, differentiation and FGFR signalling could be rescued by the addition of exogenous heparin to the media (Johnson et al. 2007; Kraushaar et al. 2010) underlining the essential role of HS in transmitting signals essential for ES cell differentiation.

5.4 HSPGs and Human ES Cells

In terms of hES cells, the primary goal is to provide differentiated cells for cell therapy. The starting point, as with mES cells, is to understand their basic biology, and many groups are working towards this. Instructive methods of efficiently differentiating hES cells generally involve adding cocktails of growth factors in a particular sequence to the differentiating culture to instruct the cells towards the required fate. It is our belief that defined HS species introduced at key stages of differentiation, perhaps in combination with protein factors could be used to improve the efficacy of differentiation regimens by mimicking the mechanisms used by cells themselves. The necessary background understanding of the role of HS in hES cell biology is at an early stage, but the FGF-2 dependency of hES cultures suggests HSPGs may have a key role to play. Indeed as discussed already, a recent publication from one of the major hES research groups identifies MEF-secreted HSPG as the key component of MEF-conditioned media that acts to stabilise FGF-2 added to the culture, in addition to directly mediating its binding to receptors on the hES cell surface (Levenstein et al. 2008).

As more information on the role of GAGs in defining cell fate becomes available, we are likely to see a dramatic increase in the attention given to the role of GAGs in mediating stem cell behaviour. With significant improvements in methods for the structural analysis of GAGs produced *in vivo*, as well as the production of synthetic oligosaccharides, the ability of saccharides to simultaneously influence multiple signalling pathways makes them key targets for the development of stem cell-based therapeutics. They are an attractive option for this as they involve neither genetic manipulation nor addition of complex, often animal-derived protein factors. As significant as heparin has been in the clinic for the treatment of thromboses, the near future may well see other saccharides playing critical roles in the development of cells for the treatment of neurological disorders, heart disease, diabetes and blindness.

References

Ai X, Do AT, Lozynska O, Kusche-Gullberg M, Lindahl U, Emerson CP Jr (2003) QSulf1 remodels the 6-O sulfation states of cell surface heparan sulfate proteoglycans to promote Wnt signaling. J Cell Biol 162:341–351

Aikawa J, Esko JD (1999) Molecular cloning and expression of a third member of the heparan sulfate/heparin GlcNAc N-deacetylase/ N-sulfotransferase family. J Biol Chem 274: 2690–2695

Aikawa J, Grobe K, Tsujimoto M, Esko JD (2001) Multiple isozymes of heparan sulfate/heparin GlcNAc N-deacetylase/GlcN N-sulfotransferase. Structure and activity of the fourth member, NDST4. J Biol Chem 276:5876–5882

Allen BL, Rapraeger AC (2003) Spatial and temporal expression of heparan sulfate in mouse development regulates FGF and FGF receptor assembly. J Cell Biol 163:637–648

Amit M, Carpenter MK, Inokuma MS, Chiu CP, Harris CP, Waknitz MA, Itskovitz-Eldor J, Thomson JA (2000) Clonally derived human embryonic stem cell lines maintain pluripotency and proliferative potential for prolonged periods of culture. Dev Biol 227:271–278

Amit M, Shariki C, Margulets V, Itskovitz-Eldor J (2004) Feeder layer- and serum-free culture of human embryonic stem cells. Biol Reprod 70:837–845

Aubert J, Stavridis MP, Tweedie S, O'Reilly M, Vierlinger K, Li M, Ghazal P, Pratt T, Mason JO, Roy D et al (2003) Screening for mammalian neural genes via fluorescence-activated cell sorter purification of neural precursors from Sox1-gfp knock-in mice. Proc Natl Acad Sci 100 (suppl 1):11836–11841

Baldwin RJ, ten Dam GB, van Kuppevelt TH, Lacaud G, Gallagher JT, Kouskoff V, Merry CL (2008) A developmentally regulated heparan sulfate epitope defines a subpopulation with increased blood potential during mesodermal differentiation. Stem Cells 26:3108–3118

Baxter MA, Camarasa MV, Bates N, Small F, Murray P, Edgar D, Kimber SJ (2009) Analysis of the distinct functions of growth factors and tissue culture substrates necessary for the long-term self-renewal of human embryonic stem cell lines. Stem Cell Res 3:28–38

Boyer LA, Lee TI, Cole MF, Johnstone SE, Levine SS, Zucker JP, Guenther MG, Kumar RM, Murray HL, Jenner RG et al (2005) Core transcriptional regulatory circuitry in human embryonic stem cells. Cell 122:947–956

Braam SR, Zeinstra L, Litjens S, Ward-van Oostwaard D, van den Brink S, van Laake L, Lebrin F, Kats P, Hochstenbach R, Passier R et al (2008) Recombinant vitronectin is a functionally defined substrate that supports human embryonic stem cell self-renewal via alphavbeta5 integrin. Stem Cells 26:2257–2265

Bullock SL, Fletcher JM, Beddington RS, Wilson VA (1998) Renal agenesis in mice homozygous for a gene trap mutation in the gene encoding heparan sulfate 2-sulfotransferase. Genes Dev 12:1894–1906

Chew JL, Loh YH, Zhang W, Chen X, Tam WL, Yeap LS, Li P, Ang YS, Lim B, Robson P et al (2005) Reciprocal transcriptional regulation of Pou5f1 and Sox2 via the Oct4/Sox2 complex in embryonic stem cells. Mol Cell Biol 25:6031–6046

Choi K, Kennedy M, Kazarov A, Papadimitriou JC, Keller G (1998) A common precursor for hematopoietic and endothelial cells. Development 125:725–732

David G, Bai XM, Van der Schueren B, Cassiman JJ, Van den Berghe H (1992) Developmental changes in heparan sulfate expression: in situ detection with mAbs. J Cell Biol 119:961–975

Dhoot GK, Gustafsson MK, Ai X, Sun W, Standiford DM, Emerson CP Jr (2001) Regulation of Wnt signaling and embryo patterning by an extracellular sulfatase. Science 293: 1663–1666

Eiselleova L, Matulka K, Kriz V, Kunova M, Schmidtova Z, Neradil J, Tichy B, Dvorakova D, Pospisilova S, Hampl A et al (2009) A complex role for FGF-2 in self-renewal, survival, and adhesion of human embryonic stem cells. Stem Cells 27:1847–1857

Esko JD, Lindahl U (2001) Molecular diversity of heparan sulfate. J Clin Invest 108:169–173

Evans MJ, Kaufman MH (1981) Establishment in culture of pluripotential cells from mouse embryos. Nature 292:154–156

Fehling HJ, Lacaud G, Kubo A, Kennedy M, Robertson S, Keller G, Kouskoff V (2003) Tracking mesoderm induction and its specification to the hemangioblast during embryonic stem cell differentiation. Development 130:4217–4227

Fong H, Hohenstein KA, Donovan PJ (2008) Regulation of self-renewal and pluripotency by Sox2 in human embryonic stem cells. Stem Cells 26:1931–1938

Ford-Perriss M, Guimond SE, Greferath U, Kita M, Grobe K, Habuchi H, Kimata K, Esko JD, Murphy M, Turnbull JE (2002) Variant heparan sulfates synthesized in developing mouse brain differentially regulate FGF signalling. Glycobiology 12:721–727

Forsberg E, Kjellen L (2001) Heparan sulfate: lessons from knockout mice. J Clin Invest 108:175–180

Forsberg E, Pejler G, Ringvall M, Lunderius C, Tomasini-Johansson B, Kusche-Gullberg M, Eriksson I, Ledin J, Hellman L, Kjellen L (1999) Abnormal mast cells in mice deficient in a heparin-synthesizing enzyme. Nature 400:773–776

Fujise M, Takeo S, Kamimura K, Matsuo T, Aigaki T, Izumi S, Nakato H (2003) Dally regulates Dpp morphogen gradient formation in the Drosophila wing 10.1242/dev.00379. Development 130:1515–1522

Furue MK, Na J, Jackson JP, Okamoto T, Jones M, Baker D, Hata R, Moore HD, Sato JD, Andrews PW (2008) Heparin promotes the growth of human embryonic stem cells in a defined serum-free medium. Proc Natl Acad Sci USA 105:13409–13414

Gallagher JT (1998) The interaction and regulation of basic and acidic fibroblast growth factors by heparan sulphate. Trends Glycosci Glycotech 10:137–144

Gallagher JT (2001) Heparan sulphate: growth control with a restricted sequence menu. J Clin Invest 108:357–361

Girardin EP, Hajmohammadi S, Birmele B, Helisch A, Shworak NW, de Agostini AI (2005) Synthesis of anticoagulantly active heparan sulfate proteoglycans by glomerular epithelial cells involves multiple 3-O-sulfotransferase isoforms and a limiting precursor pool. J Biol Chem 280:38059–38070

Grobe K, Inatani M, Pallerla SR, Castagnola J, Yamaguchi Y, Esko JD (2005) Cerebral hypoplasia and craniofacial defects in mice lacking heparan sulfate Ndst1 gene function. Development 132:3777–3786

Guimond SE, Turnbull JE (1999) Fibroblast growth factor receptor signalling is dictated by specific heparan sulphate saccharides. Curr Biol 9:1343–1346

Guimond S, Maccarana M, Olwin B, Lindahl U, Rapraeger A (1993) Activating and inhibitory heparin sequences for FGF-2 (basic FGF). Distinct requirements for FGF-1, FGF-2, and FGF-4. J Biol Chem 268:23906–23914

Habuchi H, Tanaka M, Habuchi O, Yoshida K, Suzuki H, Ban K, Kimata K (2000) The occurrence of three isoforms of heparan sulfate 6-O-sulfotransferase having different specificities for hexuronic acid adjacent to the targeted N-sulfoglucosamine. J Biol Chem 275:2859–2868

Habuchi H, Nagai N, Sugaya N, Atsumi F, Stevens RL, Kimata K (2007) Mice deficient in heparan sulfate 6-O-sulfotransferase-1 exhibit defective heparan sulfate biosynthesis, abnormal placentation, and late embryonic lethality. J Biol Chem 282:15578–15588

HajMohammadi S, Enjyoji K, Princivalle M, Christi P, Lech M, Beeler D, Rayburn H, Schwartz JJ, Barzegar S, de Agostini AI et al (2003) Normal levels of anticoagulant heparan sulfate are not essential for normal hemostasis. J Clin Invest 111:989–999

Holley RJ, Pickford CE, Rushton G, Lacaud G, Gallagher JT, Kouskoff V, Merry CL (2010) Influencing hematopoietic differentiation of mouse embryonic stem cells using soluble heparin and heparan sulfate saccharides. J Biol Chem Dec 10 2010

Holmborn K, Ledin J, Smeds E, Eriksson I, Kusche-Gullberg M, Kjellen L (2004) Heparan sulfate synthesized by mouse embryonic stem cells deficient in NDST1 and NDST2 is 6-O-sulfated but contains no N-sulfate groups. J Biol Chem 279:42355–42358

Holst CR, Bou-Reslan H, Gore BB, Wong K, Grant D, Chalasani S, Carano RA, Frantz GD, Tessier-Lavigne M, Bolon B et al (2007) Secreted sulfatases Sulf1 and Sulf2 have overlapping yet essential roles in mouse neonatal survival. PLoS ONE 2:e575

Inatani M, Yamaguchi Y (2003) Gene expression of EXT1 and EXT2 during mouse brain development. Dev Brain Res 141:129–136

Johnson CE, Crawford BE, Stavridis M, Ten Dam G, Wat AL, Rushton G, Ward CM, Wilson V, van Kuppevelt TH, Esko JD et al (2007) Essential alterations of heparan sulfate during the differentiation of embryonic stem cells to Sox1-enhanced green fluorescent protein-expressing neural progenitor cells. Stem Cells 25:1913–1923

Kennedy M, Firpo M, Choi K, Wall C, Robertson S, Kabrun N, Keller G (1997) A common precursor for primitive erythropoiesis and definitive haematopoiesis. Nature 386:488–493

Kleinman HK, McGarvey ML, Liotta LA, Robey PG, Tryggvason K, Martin GR (1982) Isolation and characterization of type IV procollagen, laminin, and heparan sulfate proteoglycan from the EHS sarcoma. Biochemistry 21:6188–6193

Kraushaar DC, Yamaguchi Y, Wang L (2010) Heparan sulfate is required for embryonic stem cells to exit from self-renewal. J Biol Chem 285:5907–5916

Kreuger J, Jemth P, Sanders-Lindberg E, Eliahu L, Ron D, Basilico C, Salmivirta M, Lindahl U (2005) Fibroblast growth factors share binding sites in heparan sulphate. Biochem J 389: 145–150

Lamanna WC, Baldwin RJ, Padva M, Kalus I, Ten Dam G, van Kuppevelt TH, Gallagher JT, von Figura K, Dierks T, Merry CL (2006) Heparan sulfate 6-O-endosulfatases: discrete in vivo activities and functional co-operativity. Biochem J 400:63–73

Lander AD (2005) Heparan sulfate and the molecular teleology of co-reception. In: Proteoglycans in signalling. Uppsala, Sweden

Lanner F, Lee KL, Sohl M, Holmborn K, Yang H, Wilbertz J, Poellinger L, Rossant J, Farnebo F (2009) Heparan sulfation dependent FGF signalling maintains embryonic stem cells primed for differentiation in a heterogeneous state. Stem Cells 28:191–200

Lecuit T, Cohen S (1998) Dpp receptor levels contribute to shaping the Dpp morphogen gradient in the Drosophila wing imaginal disc. Development 125:4901–4907

Ledin J, Staatz W, Li JP, Gotte M, Selleck S, Kjellen L, Spillmann D (2004) Heparan sulfate structure in mice with genetically modified heparan sulfate production. J Biol Chem 279: 42732–42741

Levenstein ME, Ludwig TE, Xu RH, Llanas RA, VanDenHeuvel-Kramer K, Manning D, Thomson JA (2006) Basic fibroblast growth factor support of human embryonic stem cell self-renewal. Stem Cells 24:568–574

Levenstein ME, Berggren WT, Lee JE, Conard KR, Llanas RA, Wagner RJ, Smith LM, Thomson JA (2008) Secreted proteoglycans directly mediate human embryonic stem cell-basic fibroblast growth factor 2 interactions critical for proliferation. Stem Cells 26:3099–3107

Li JP, Gong F, Hagner-McWhirter A, Forsberg E, Abrink M, Kisilevsky R, Zhang X, Lindahl U (2003) Targeted disruption of a murine glucuronyl C5-epimerase gene results in heparan sulfate lacking L-iduronic acid and in neonatal lethality. J Biol Chem 278:28363–28366

Li Y, Powell S, Brunette E, Lebkowski J, Mandalam R (2005) Expansion of human embryonic stem cells in defined serum-free medium devoid of animal-derived products. Biotechnol Bioeng 91:688–698

Lin X, Wei G, Shi Z, Dryer L, Esko JD, Wells DE, Matzuk MM (2000) Disruption of gastrulation and heparan sulfate biosynthesis in EXT1-deficient mice. Dev Biol 224:299–311

Lindahl U, Bäckström G, Thunberg L, Leder IG (1980) Evidence for a 3-O-sulfated D-glucosamine residue in the antithrombin-binding sequence of heparin. Proc Natl Acad Sci USA 77: 6551–6555

Loh YH, Wu Q, Chew JL, Vega VB, Zhang W, Chen X, Bourque G, George J, Leong B, Liu J et al (2006) The Oct4 and Nanog transcription network regulates pluripotency in mouse embryonic stem cells. Nat Genet 38:431–440

Lu J, Hou R, Booth CJ, Yang SH, Snyder M (2006) Defined culture conditions of human embryonic stem cells. Proc Natl Acad Sci USA 103:5688–5693

Ludwig TE, Bergendahl V, Levenstein ME, Yu J, Probasco MD, Thomson JA (2006a) Feeder-independent culture of human embryonic stem cells. Nat Methods 3:637–646

Ludwig TE, Levenstein ME, Jones JM, Berggren WT, Mitchen ER, Frane JL, Crandall LJ, Daigh CA, Conard KR, Piekarczyk MS et al (2006b) Derivation of human embryonic stem cells in defined conditions. Nat Biotechnol 24:185–187

Lum DH, Tan J, Rosen SD, Werb Z (2007) Gene trap disruption of the mouse heparan sulfate 6-O-endosulfatase gene, Sulf2. Mol Cell Biol 27:678–688

Lyon M, Gallagher JT (1998) Bio-specific sequences and domains in heparan sulphate and the regulation of cell growth and adhesion [In Process Citation]. Matrix Biol 17:485–493

Maccarana M, Sakura Y, Tawada A, Yoshida K, Lindahl U (1996) Domain structure of heparan sulfates from bovine organs. J Biol Chem 271:17804–17810

Martin GR (1981) Isolation of a pluripotent cell line from early mouse embryos cultured in medium conditioned by teratocarcinoma stem cells. Proc Natl Acad Sci USA 78:7634–7638

Martin MJ, Muotri A, Gage F, Varki A (2005) Human embryonic stem cells express an immunogenic nonhuman sialic acid. Nat Med 11:228–232

Merry CL, Bullock SL, Swan DC, Backen AC, Lyon M, Beddington RS, Wilson VA, Gallagher JT (2001) The molecular phenotype of heparan sulfate in the Hs2st-/- mutant mouse. J Biol Chem 276:35429–35434

Miosge N, Simniok T, Sprysch P, Herken R (2003) The collagen Type XVIII endostatin domain is co-localized with perlecan in basement membranes in vivo. J Histochem Cytochem 51: 285–296

Mitsui K, Tokuzawa Y, Itoh H, Segawa K, Murakami M, Takahashi K, Maruyama M, Maeda M, Yamanaka S (2003) The homeoprotein Nanog is required for maintenance of pluripotency in mouse epiblast and ES cells. Cell 113:631–642

Murphy KJ, Merry CL, Lyon M, Thompson JE, Roberts IS, Gallagher JT (2004) A new model for the domain structure of heparan sulfate based on the novel specificity of K5 lyase. J Biol Chem 279:27239–27245

Niwa H (2007) How is pluripotency determined and maintained? Development 134:635–646

Niwa H, Miyazaki J, Smith AG (2000) Quantitative expression of Oct-3/4 defines differentiation, dedifferentiation or self-renewal of ES cells. Nat Genet 24:372–376

Niwa H, Toyooka Y, Shimosato D, Strumpf D, Takahashi K, Yagi R, Rossant J (2005) Interaction between Oct3/4 and Cdx2 determines trophectoderm differentiation. Cell 123:917–929

Noonan D, Fulle A, Valente P, Cai S, Horigan E, Sasaki M, Yamada Y, Hassell J (1991) The complete sequence of perlecan, a basement membrane heparan sulfate proteoglycan, reveals extensive similarity with laminin A chain, low density lipoprotein-receptor, and the neural cell adhesion molecule. J Biol Chem 266:22939–22947

Paine-Saunders S, Viviano BL, Economides AN, Saunders S (2002) Heparan sulfate proteoglycans retain Noggin at the cell surface: a potential mechanism for shaping bone morphogenetic protein gradients. J Biol Chem 277:2089–2096

Pallerla SR, Lawrence R, Lewejohann L, Pan Y, Fischer T, Schlomann U, Zhang X, Esko JD, Grobe K (2008) Altered heparan sulfate structure in mice with deleted NDST3 gene function. J Biol Chem 283:16885–16894

Pickford CE, Holley RJ, Rushton G, Stavridis MP, Ward CM, Merry CL (2011) Specific Glycosaminoglycans Modulate Neural Specification of Mouse Embryonic Stem Cells. Stem Cells Feb 4

Pratt T, Conway CD, Tian NM, Price DJ, Mason JO (2006) Heparan sulphation patterns generated by specific heparan sulfotransferase enzymes direct distinct aspects of retinal axon guidance at the optic chiasm. J Neurosci 26:6911–6923

Pye DA, Vives RR, Turnbull JE, Hyde P, Gallagher JT (1998) Heparan sulfate oligosaccharides require 6-O-sulfation for promotion of basic fibroblast growth factor mitogenic activity. J Biol Chem 273:22936–22942

Qi X, Li TG, Hao J, Hu J, Wang J, Simmons H, Miura S, Mishina Y, Zhao GQ (2004) BMP4 supports self-renewal of embryonic stem cells by inhibiting mitogen-activated protein kinase pathways. Proc Natl Acad Sci USA 101:6027–6032

Rapraeger AC, Guimond S, Krufka A, Olwin BB (1994) Regulation by heparan sulfate in fibroblast growth factor signaling. Methods Enzymol 245:219–240

Ratzka A, Kalus I, Moser M, Dierks T, Mundlos S, Vortkamp A (2008) Redundant function of the heparan sulfate 6-O-endosulfatases Sulf1 and Sulf2 during skeletal development. Dev Dyn 237:339–353

Ringvall M, Ledin J, Holmborn K, van Kuppevelt T, Ellin F, Eriksson I, Olofsson AM, Kjellen L, Forsberg E (2000) Defective heparan sulfate biosynthesis and neonatal lethality in mice lacking N-deacetylase/N-sulfotransferase-1. J Biol Chem 275:25926–25930

Robinson CJ, Mulloy B, Gallagher JT, Stringer SE (2006) VEGF165-binding sites within heparan sulfate encompass two highly sulfated domains and can be liberated by K5 lyase. J Biol Chem 281:1731–1740

Rodda DJ, Chew JL, Lim LH, Loh YH, Wang B, Ng HH, Robson P (2005) Transcriptional regulation of nanog by OCT4 and SOX2. J Biol Chem 280:24731–24737

Sasaki N, Okishio K, Ui-Tei K, Saigo K, Kinoshita-Toyoda A, Toyoda H, Nishimura T, Suda Y, Hayasaka M, Hanaoka K et al (2008) Heparan sulfate regulates self-renewal and pluripotency of embryonic stem cells. J Biol Chem 283:3594–3606

Sedita J, Izvolsky K, Cardoso WV (2004) Differential expression of heparan sulfate 6-O-sulfotransferase isoforms in the mouse embryo suggests distinctive roles during organogenesis. Dev Dyn 231:782–794

Shworak NW, Fritze LMS, Liu J, Butler LD, Rosenberg RD (1996) Cell-free synthesis of anticoagulant heparan sulfate reveals a limiting converting activity that modifies an excess precursor pool. J Biol Chem 271:27063–27071

Shworak NW, Liu J, Petros LM, Zhang L, Kobayashi M, Copeland NG, Jenkins NA, Rosenberg RD (1999) Multiple isoforms of heparan sulfate D-glucosaminyl 3-O-sulfotransferase. Isolation, characterization, and expression of human cdnas and identification of distinct genomic loci. J Biol Chem 274:5170–5184

Shworak NW, HajMohammadi S, de Agostini AI, Rosenberg RD (2002) Mice deficient in heparan sulfate 3-O-sulfotransferase-1: normal hemostasis with unexpected perinatal phenotypes. Glycoconj J 19:355–361

Smith A, Heath J, Donaldson D, Wong G, Moreau J, Stahl M, Rogers D (1988) Inhibition of pluripotential embryonic stem cell differentiation by purified polypeptides. Nature 336: 688–690

Smith RA, Meade K, Pickford CE, Holley RJ, Merry CL (2011) Glycosaminoglycans as regulators of stem cell differentiation. Biochem Soc Trans 39:383–387

Soker S, Goldstaub D, Svahn CM, Vlodavsky I, Levi BZ, Neufeld G (1994) Variations in the size and sulfation of heparin modulate the effect of heparin on the binding of VEGF165 to its receptors. Biochem Biophys Res Commun 203:1339–1347

Stavridis MP, Lunn JS, Collins BJ, Storey KG (2007) A discrete period of FGF-induced Erk1/2 signalling is required for vertebrate neural specification. Development 134:2889–2894

Stavridis MP, Collins BJ, Storey KG (2010) Retinoic acid orchestrates fibroblast growth factor signalling to drive embryonic stem cell differentiation. Development 137:881–890

Stewart R, Stojkovic M, Lako M (2006) Mechanisms of self-renewal in human embryonic stem cells. Eur J Cancer 42:1257–1272

Stickens D, Zak BM, Rougier N, Esko JD, Werb Z (2005) Mice deficient in Ext2 lack heparan sulfate and develop exostoses. Development 132:5055–5068

Tam PP, Behringer RR (1997) Mouse gastrulation: the formation of a mammalian body plan. Mech Dev 68:3–25

ten Dam GB, Hafmans T, Veerkamp JH, van Kuppevelt TH (2003) Differential expression of heparan sulfate domains in rat spleen. J Histochem Cytochem 51:727–739

Terman B, Khandke L, Dougher-Vermazan M, Maglione D, Lassam NJ, Gospodarowicz D, Persico MG, Bohlen P, Eisinger M (1994) VEGF receptor subtypes KDR and FLT1 show different sensitivities to heparin and placenta growth factor. Growth Factors 11:187–195

Tessler S, Rockwell P, Hicklin D, Cohen T, Levi BZ, Witte L, Lemischka IR, Neufeld G (1994) Heparin modulates the interaction of VEGF165 with soluble and cell associated flk-1 receptors. J Biol Chem 269:12456–12461

Thomson JA, Itskovitz-Eldor J, Shapiro SS, Waknitz MA, Swiergiel JJ, Marshall VS, Jones JM (1998) Embryonic stem cell lines derived from human blastocysts. Science 282:1145–1147

Tsen G, Halfter W, Kröger S, Cole GJ (1995) Agrin is a heparan sulfate proteoglycan. J Biol Chem 270:3392–3399

Turnbull JE, Gallagher JT (1993) Heparan sulphate: functional role as a modulator of fibroblast growth factor activity. Biochem Soc Trans 21:477–482

Turnbull JE, Fernig DG, Ke Y, Wilkinson MC, Gallagher JT (1992) Identification of the basic fibroblast growth factor binding sequence in fibroblast heparan sulfate. J Biol Chem 267: 10337–10341

van Kuppevelt TH, Dennissen MA, van Venrooij WJ, Hoet RM, Veerkamp JH (1998) Generation and application of type-specific anti-heparan sulfate antibodies using phage display technology. Further evidence for heparan sulfate heterogeneity in the kidney. J Biol Chem 273: 12960–12966

Wang S, Ai X, Freeman SD, Pownall ME, Lu Q, Kessler DS, Emerson CP Jr (2004) QSulf1, a heparan sulfate 6-O-endosulfatase, inhibits fibroblast growth factor signaling in mesoderm induction and angiogenesis. Proc Natl Acad Sci USA 101:4833–4838

Ward C, Stern P, Willington M, Flenniken A (2002) Efficient germline transmission of mouse embryonic stem cells grown in synthetic serum in the absence of a fibroblast feeder layer. Lab Invest 82:1765–1767

Williams RL, Hilton DJ, Pease S, Willson TA, Stewart CL, Gearing DP, Wagner EF, Metcalf D, Nicola NA, Gough NM (1988) Myeloid leukaemia inhibitory factor maintains the developmental potential of embryonic stem cells. Nature 336:684–687

Wood HB, Episkopou V (1999) Comparative expression of the mouse Sox1, Sox2 and Sox3 genes from pre-gastrulation to early somite stages. Mech Dev 86:197–201

Xu C, Inokuma MS, Denham J, Golds K, Kundu P, Gold JD, Carpenter MK (2001) Feeder-free growth of undifferentiated human embryonic stem cells. Nat Biotechnol 19:971–974

Xu C, Rosler E, Jiang J, Lebkowski JS, Gold JD, O'Sullivan C, Delavan-Boorsma K, Mok M, Bronstein A, Carpenter MK (2005a) Basic fibroblast growth factor supports undifferentiated human embryonic stem cell growth without conditioned medium. Stem Cells 23:315–323

Xu RH, Peck RM, Li DS, Feng X, Ludwig T, Thomson JA (2005b) Basic FGF and suppression of BMP signaling sustain undifferentiated proliferation of human ES cells. Nat Methods 2: 185–190

Yabe T, Hata T, He J, Maeda N (2005) Developmental and regional expression of heparan sulfate sulfotransferase genes in the mouse brain. Glycobiology 15:982–993

Ying QL, Nichols J, Chambers I, Smith A (2003a) BMP induction of Id proteins suppresses differentiation and sustains embryonic stem cell self-renewal in collaboration with STAT3. Cell 115:281–292

Ying QL, Stavridis M, Griffiths D, Li M, Smith A (2003b) Conversion of embryonic stem cells into neuroectodermal precursors in adherent monoculture. Nat Biotechnol 21:183–186

Ying QL, Wray J, Nichols J, Batlle-Morera L, Doble B, Woodgett J, Cohen P, Smith A (2008) The ground state of embryonic stem cell self-renewal. Nature 453:519–523

Zhao W, Sala-Newby GB, Dhoot GK (2006) Sulf1 expression pattern and its role in cartilage and joint development. Dev Dyn 235:3327–3335

Chapter 6
Heparan Sulfate Proteoglycans in Cancer Therapy

Ezeddin Salem Gassar, Sherif A. Ibrahim, and Martin Götte

Abstract The polyanionic linear polysaccharide heparan sulfate specifically interacts with a multitude of extracellular ligands relevant to all steps of tumor progression. Heparan sulfate proteoglycans act as coreceptors for cytokine and chemokine signaling, modulating tumor cell growth and survival, chemotaxis, and angiogenesis. As matrix receptors, they act in concert with integrins to regulate tumor cell motility. As binding partners for matrix metalloproteinases and protease inhibitors, they regulate the proteolytic microenvironment of tumors, thus modulating metastatic spread. Processing enzymes such as heparanase and HSulf-1 and HSulf-2 can further modify the biochemical properties of heparan sulfate and promote tumor progression. As dysregulated expression of heparan sulfate proteoglycans and heparan sulfate-processing enzymes has been observed in numerous tumor entities, the development of glycan-based drugs targeting their biological functions has become an area of intense research. Promising results have been obtained both in animal models and in phase I–III clinical trials.

6.1 Introduction

Heparan sulfate (HS) is an unbranched polysaccharide composed of repetitive disaccharide units which consists of alternating α-D-glucosamine residues linked to either α-L-iduronic acid or β-D-glucuronic acid (Bernfield et al. 1999; Kusche-Gullberg and Kjellen 2003). The regulated introduction or removal of N-acetylation, or of sulfate groups at the N-, 2-O, 6-O, and 3-O positions of the carbohydrate chain, as well as epimerization of the glucuronic acid, results in a high degree of

E.S. Gassar, S.A. Ibrahim, and M. Götte (✉)
Department of Gynecology and Obstetrics, Münster University Hospital, Albert-Schweitzer-Strasse 33, 48149 Münster, Germany
e-mail: mgotte@uni-muenster.de

M.S.G. Pavão (ed.), *Glycans in Diseases and Therapeutics*, Biology of Extracellular Matrix,
DOI 10.1007/978-3-642-16833-8_6,

structural diversity and enables HS to form specific binding motifs for a large variety of ligands relevant to cancer progression (Yip et al. 2006; Casu et al. 2008; Borsig 2010). Via covalent attachment to core proteins, HS forms proteoglycans (HSPGs), utilizing an O-glycosidic linkage formed by the tetrasaccharide glucuronyl-β-(1-3)-galactosyl-β-(1-3)-galactosyl-β-(1-4)-xylosyl-β-1-O-(serine). For a more detailed description of HS biosynthesis, its structural domain organization, and the basic biological functions of HSPGs, the reader is referred to Chaps. 2 and 5.

HSPGs are ubiquitously expressed at the cell surface and – as secreted proteoglycans or shed ectodomains of cell surface HSPGs – in basement membranes and extracellular matrices (Bernfield et al. 1999; Iozzo et al. 2009). Most of the molecular functions of HSPGs are mediated by specific binding actions of their HS chains with growth factors and chemokines, their cell surface receptors, enzymes, and enzyme inhibitors, as well as extracellular matrix proteins (Bernfield et al. 1999; Iozzo et al. 2009). While this chapter focuses on the HS-mediated functions of HSPGs in cancer, it has to be considered that some of the functions of HSPGs may be mediated via direct binding interactions of ligands to the core protein, that some HSPGs can be substituted with more than one type of glycosaminoglycan chain, and that some HSPGs only constitute part-time proteoglycans (Bernfield et al. 1999; Yip et al. 2006; Götte and Yip 2006). Furthermore, most HSPGs contain N-glycan substitutions in addition to their HS chains, which may contribute to their molecular role in tumor progression (cf Chap. 3).

The strategic placement of HSPGs at the interface between the cell and its surrounding extracellular matrix environment suggests that they might be critical modulators of cancer onset and progression. Depending on the structural organization and thus the ligand binding capacity of the HS chain, and on the location of its expression (tumor versus stroma, cell surface versus basement membrane, or interstitial matrix), HSPGs play different, and sometimes opposing, roles in tumor initiation, progression, and metastasis (Yip et al. 2006; Fuster and Esko 2005). Tumor progression requires sequential steps involving proper coordination of cell proliferation, survival, adhesion, migration, and angiogenesis. In fact, numerous studies have specified roles for HSPGs as modulators in each step of tumor progression (Fig. 6.1). In a variety of tumor entities, prognostic or predictive values have been established for the expression of HSPGs and HS-biosynthetic and HS-processing enzymes, including Syndecan-1 (Sdc-1), Sdc-2, glypican-1, glypican-3, perlecan, and heparanase (cf Yip et al. 2006 for review). Moreover, mutations in the genes encoding the heparan sulfate copolymerases EXT1 and EXT2 are causative for the development of hereditary multiple exostoses and have been linked to a predeposition for developing malignant chondrosarcomas (Senay et al. 2000). In the following sections, we delineate the role of dysregulated molecular functions of HSPGs in tumor progression. Furthermore, we present HSPG-centered concepts of cancer therapy.

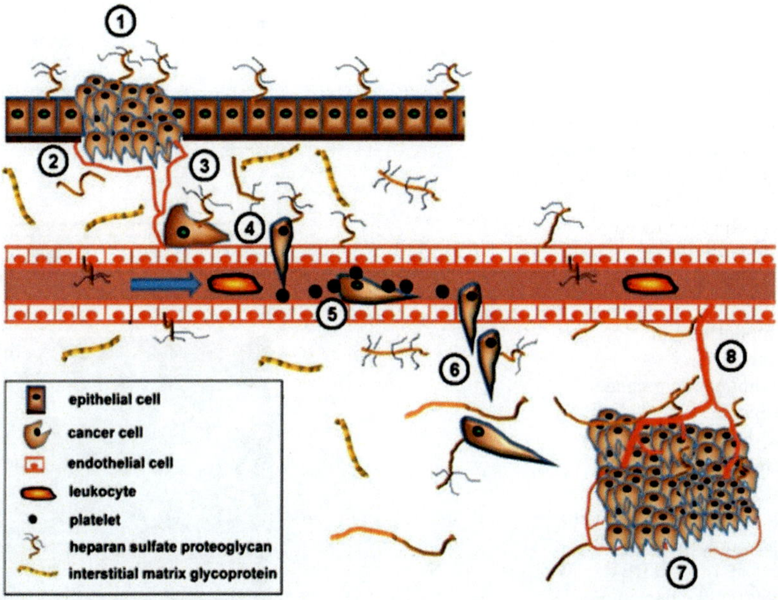

Fig. 6.1 HSPGs contribute to several key events in malignant transformation and tumor progression (1) by modulating mitogenic signaling and cell survival, leading to enhanced primary tumor growth, and by regulating epithelial-to-mesenchymal transition, resulting in loss of contact inhibition and acquisition of a migratory phenotype; (2) by modulating MMP and heparanase expression and activation, promoting degradation of basement membrane proteins and proteoglycans; (3) by modulating tumor angiogenesis; (4) by regulating chemokine functions, promoting intravasation; (5) by modulating circulating tumor cell adhesion, platelet aggregation and coagulation; (6) via regulation of endothelial adhesion, chemokine activity, cell motility, and proteolysis during tumor cell extravasation; (7) by promoting mitogenic signaling and cell survival in metastasized tumor cells; and (8) by promoting angiogenesis in distant metastases

6.2 Contribution of Dysregulated Molecular Functions of HSPGs to Tumor Progression

6.2.1 Heparan Sulfate-Dependent Regulation of Tumor Growth

6.2.1.1 Heparan Sulfate as a Modulator of Tumor Cell Proliferation

The high proliferation rate and unrestricted proliferative potential of tumor cells is a major characteristic of malignant transformation. A role for HSPGs in controlling cell proliferation is well established, as cell surface HSPGs serve as coreceptors for several growth factor tyrosine kinase receptors (Bernfield et al. 1999; Spivak-Kroizman et al. 1994) (Fig. 6.2). According to the coreceptor model, mitogenic growth factor signaling is promoted by the formation of a ternary complex of ligand, receptor, and HSPG (Schlessinger 2000). Consequently, HS on breast

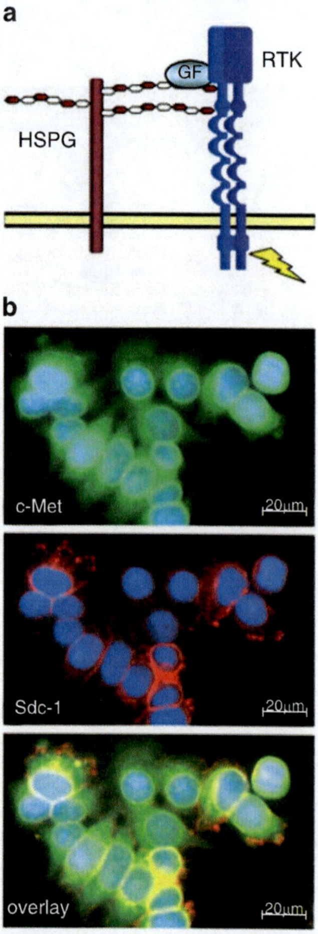

Fig. 6.2 Cell surface HSPG as a coreceptor for receptor tyrosine kinase (RTK)-mediated mitogenic signaling. (**a**) The HS chains form a ternary complex with the growth factor (GF) and the RTK, thus facilitating activation of the receptor and downstream signaling in the context of tumor cell proliferation, angiogenesis, and modulation of cell motility. (**b**) Immunofluorescence micrograph demonstrating colocalization of the hepatocyte growth factor (HGF) receptor tyrosin kinase cMet (mAb C-12, Santa Cruz) and its coreceptor HSPG syndecan-1 (mAb BB-4, Serotec) in the human breast cancer cell line MCF-7. Photograph courtesy of M. Götte and V. Samailova

cancer cells regulates the differential mitogenic activity of fibroblast growth factor (FGF)-1 and FGF-2 (Fernig et al. 2000). Concerning specific HSPGs, glypican-1 was shown to be overexpressed in breast cancer and to modulate response to FGF-2 and heparin-binding epidermal growth factor (HB-EGF)-like growth factor (Matsuda et al. 2001). siRNA-mediated knockdown of the HSPG syndecan-1 (Sdc-1) results in reduced FGF-2-dependent activation of the mitogen-activated protein kinase (MAPK) pathway in MCF-7 breast cancer and Caco-2 colon carcinoma cells (Nikolova et al. 2009; Floer et al. 2010). Sdc-1 also modulates multiple myeloma cell responses to epidermal growth factor (EGF)-family ligands of ErbB receptors (Mahtouk et al. 2006). Furthermore, a coreceptor role for Sdc-1 in HGF-mediated mitogenic and angiogenic signaling via the c-Met RTK has been established in multiple myeloma (Derksen et al. 2002) and breast cancer cells (Götte et al. 2010). Of note, soluble and membrane-bound forms of Sdc-1 appear to have opposing effects on breast cancer cell proliferation, as overexpression of the soluble form of Sdc-1 inhibited proliferation of MCF-7 breast cancer cells, possibly due to competitive binding to growth factors (Nikolova et al. 2009). However, the source (and thus, the HS fine structure) of soluble Sdc-1 also appears to influence its impact on tumor cell proliferation, as stroma-cell-derived Sdc-1 exerts an HS- and FGF-2/stroma-cell-derived factor (SDF)-1-dependent stimulatory effect on T47D breast cancer cell proliferation (Maeda et al. 2004; Su et al. 2007). Moreover, the activity of growth factors bound to HSPGs in some instances depends on their release by degradative or editing enzymes such as heparanase or HSulf-1 (Kato et al. 1998; Lai et al. 2003). Thus, although a role for HSPGs in modulating cancer cell and tumor growth is well established, their mode and direction of action are highly context-dependent.

6.2.1.2 Heparan Sulfate as a Modulator of Tumor Cell Apoptosis

Acquired or inherent resistance to apoptosis, the mechanism of programmed cell death, is a hallmark of human cancers (Fulda 2009). HSPGs modulate tumor cell apoptosis (resistance) via direct interference with apoptotic pathways, and by altering the susceptibility to apoptosis-inducing anticancer therapies (cf Sect. 1.3.3). For example, reexpression of the HS endosulfatase HSulf-1 enhanced sensitivity of ovarian cancer and hepatocellular carcinoma cell lines and xenografts to apoptosis induction by staurosporine and cisplatin, which was mechanistically related to decreased heparin-binding growth factor signaling (Lai et al. 2003, 2004a, 2006). In addition, an apoptosis-modulating role of heparanase has been described in different tumor entities (Fux et al. 2009; Ikeguchi et al. 2003; Cohen et al. 2006). Among the HSPGs regulating tumor cell apoptosis, members of the syndecan family have emerged as prime candidates. The addition of purified Sdc-1 ectodomain leads to apoptosis induction, dose-dependent growth inhibition, and concurrent down-regulation of cyclin D1 in multiple myeloma cell lines (Dhodapkar et al. 1998). Similarly, siRNA-mediated knockdown of Sdc-1 expression in human multiple myeloma cells leads to reduced growth rates and a strong increase in apoptosis (Khotskaya et al. 2009; Reijmers et al. 2010). Of note, similar effects were observed upon knockdown of the HS copolymerase EXT1. Furthermore, aggregation and apoptosis induction of HT58 lymphoma cells depend on Sdc-1 and HS (Sebestyén et al. 2000). Similarly, siRNA knockdown of Sdc-1 in androgen-independent prostate cancer cell lines leads to a downregulation of NADPH oxidase 2 and induction of apoptosis (Shimada et al. 2009). In urothelial carcinoma cell lines, Sdc-1 siRNA silencing leads to a downregulation of the transcription factor junB and the long isoform of FLICE-inhibitory protein, resulting in an induction of apoptosis that could be blocked by pan-caspase inhibitors (Shimada et al. 2010). At least some of the apoptosis-related functions of Sdc-1 appear to depend on its core protein, as association of Sdc-1 with the cytoplasmic domain of $\alpha6\beta4$ integrin regulated Akt-dependent antiapoptotic pathways in a human squamous carcinoma cell line (Wang et al. 2010). Apart from Sdc-1, Sdc-2 was shown to modulate osteosarcoma cell apoptosis in a JNK-pathway-dependent manner (Modrowski et al. 2005). Furthermore, Sdc-2 sensitizes osteosarcoma cells to chemotherapy-induced apoptosis by increasing the response to methotrexate, doxorubicin, and cisplatin (Orosco et al. 2007). Finally, interactions of the integrin ligand CCN1 with both $\alpha v\beta 3$ and $\alpha 6\beta 4$ integrins and Sdc-4 significantly enhanced TRAIL-induced apoptosis in prostate carcinoma cell lines (Franzen et al. 2009). While a role for regulating tumor cell apoptosis, and for modulating chemotherapy-induced apoptotic responses in vitro and in vivo, has also been established for the HSPG glypican-3, this function may be independent of its substitution with HS chains (Gonzalez et al. 1998; Kim et al. 2003; Peters et al. 2003; Umezu et al. 2010).

6.2.2 Invasion and Metastasis of Tumor Cells

6.2.2.1 HSPG-Dependent Modulation of Cell Motility

The acquisition of migratory and invasive capabilities by tumor cells is crucial in the progression of a solid, local primary tumor to metastasis. In the first step, transformed cells within the primary tumor need to loosen cell–cell and cell–matrix contacts to migrate through the surrounding tissue. This process involves changes in the expression of cell surface molecules and frequently involves an aberrant reactivation of developmental processes such as epithelial-to-mesenchymal transition (Egeblad and Werb 2002; Micalizzi et al. 2010). Of note, many of these processes are regulated by HSPG-dependent signaling pathways such as the Wnt or notch pathways (cf Chap. 5) (Dieudonné et al. 2010; O'Connell et al. 2009; Luyten et al. 2008; Liu et al. 2003; Kamimura et al. 2004). For example, the soluble form of glypican-3 can stimulate autocrine/paracrine canonical Wnt signaling in hepatocellular carcinoma cells (Zittermann et al. 2010), and shRNA-mediated knockdown of the HS editing enzyme Sulf-2 in nonsmall cell lung cancer cells eliminated autocrine Wnt signaling and strongly inhibited xenograft tumor formation in nude mice (Lemjabbar-Alaoui et al. 2010). Furthermore, resistance of Sdc-1-deficient mice to experimentally induced tumorigenesis has been linked to an alteration in Wnt-responsive precursor cell pools (Liu et al. 2003; McDermott et al. 2007), demonstrating the importance of the role of HS(PG) in (cancer) stem cell function (cf Chap. 5). Moreover, expression of the homotypic cell adhesion molecule E-cadherin, which is frequently downregulated in invasive cancers, appears to be linked to the expression of cell surface HSPGs (Götte et al. 2007). Synchronous regulation of Sdc-1 and E-cadherin has been described in colon carcinoma, prostate cancer, and in breast cancer cell lines (Bernfield et al. 1999; Nikolova et al. 2009; Contreras et al. 2010), and similar observations have been made for glypican-3 (Peters et al. 2003). In fact, Sdc-1 expression is necessary for maintaining epithelial morphology, and when it is lost, the epithelial cell becomes fusiform and fibroblastic in appearance and exhibits characteristics of neoplastically transformed cells including anchorage-independent growth, presence of elongated filopodial extensions, and absence of stress fibers and focal adhesions (reviewed in Bernfield et al. 1999). Low levels of cell surface heparan sulfate correlate with high metastatic activity of many tumors, including colon, mesothelioma, lung, hepatocellular carcinoma, infiltrating ductal carcinoma of the breast, and head and neck carcinoma (reviewed in Yip et al.

2006). Upon acquiring a migratory phenotype, differential adhesion to interstitial ECM molecules, activation of integrins, and focal adhesion kinase signaling pathways lead to Rho-GTPase-dependent rearrangements of the cytoskeleton in tumor cells. Apparently, these processes are modulated by HSPGs (Morgan et al. 2007). For example, in squamous carcinoma cells, Sdc-1 silencing leads to reduced focal adhesion formation and enhanced cell spreading and motility on collagen I substrates (Ishikawa and Kramer 2010). While adhesion-induced RhoA activation was ablated in the Sdc-1-deficient cells, Rac1 was strongly activated, leading the authors to suggest that Sdc1 may mediate the link between integrin-induced actin remodeling and motility. In contrast, in endometrial carcinoma cells, Sdc-1 overexpression conferred a more mobile and invasive phenotype, possibly via regulating its association with $\alpha v\beta 5$ integrin and focal adhesion kinase-mediated signaling (Oh et al. 2009). Sdc-1 is also an important receptor for laminin $\alpha 1$-stimulated oral squamous cell carcinoma cell migration (Siqueira et al. 2010) and regulates $aV\beta 3$ integrin-dependent breast cancer cell adhesion and motility (Beauvais et al. 2004). In addition to ECM-induced triggering of syndecan-modulated integrin function, the coreceptor role for HSPG in chemokine signaling (cf. Chap. 1) may promote directional migration of tumor cells, as was recently shown for a HS-dependent activation of Sdc-4 by SDF-1 in a human hepatoma cell line (Sutton et al. 2007). In addition, a modulation of proinvasive RTK-mediated signaling by HSPG coreceptors, as e.g., exemplified by the HGF receptor c-Met (Fig. 6.2) contributes increased tumor cell migration and invasiveness (Götte et al. 2007, 2010). HSPGs in the extracellular matrix may further promote this process by sequestering chemokines and growth factors within close proximity to migrating tumor cells. Finally, it has to be considered that the fine structure of the HS chain of defined core proteins can also have a considerable impact on the metastatic behavior of tumor cells. For example, HS undergoes specific structural changes during the progression from human colon adenoma to carcinoma in vitro, with colon carcinoma cells displaying a 33% reduction in 2-O-sulfation on iduronic acid and a 20% reduction in overall N-sulfation compared to adenoma cells (Jayson et al. 1998). Moreover, reactivation of epigenetically silenced heparan sulfate 3-O-sulfotransferases was found to repress the invasive phenotype of a chondrosarcoma cell line (Bui et al. 2010), whereas overexpression of the HS editing enzyme HSulf-1, which regulates the 6-O-sulfation state of HS, increased the invasiveness of pancreatic and head and neck squamous carcinoma cell lines, in part via modulation of the HGF-/c-Met signaling pathway (Abiatari et al. 2006; Lai et al. 2004b).

6.2.2.2 HSPGs as Targets and Regulators of Protease and Heparanase Activity

In addition to a loss of cell–cell contact and to increased motility, tumor cells need to be able to penetrate the extracellular matrix and basement membranes in order to form distant metastases (Fig. 6.1). As HSPGs are integral constituents of basement membranes and interstitial matrices in addition to structural glycoproteins such as collagens and laminins, a tumor cell may increase its invasiveness by downregulating basement membrane HSPG expression. Consistent with this concept, reduced expression of the basement membrane HSPG perlecan was shown to be associated with tumor progression in endometrial cancer and with pelvic lymph node metastasis in invasive cervical cancer (Hasengaowa et al. 2005; Kodama et al. 2005). An additional efficient strategy of tumor cells to promote invasiveness is to utilize matrix metalloproteinases (MMPs) to degrade the surrounding extracellular matrix (Egeblad and Werb 2002). There is ample evidence for a HSPG-dependent regulation of MMP expression: Crosstalk between Sdc-1 and α2β1 integrin signaling enhances MMP-1 transcription in response to collagen binding (Vuoriluoto et al. 2008), and overexpression of Sdc-1 in endometrial carcinoma cells results in an NFkB-dependent upregulation of MMP-9 expression (Oh et al. 2009). Furthermore, interactions of Sdc-1 and β1 integrins with the laminin α1-derived peptide AG73 in salivary gland and oral squamous cell carcinoma cell lines modulate MMP9 expression and activity (Gama-de-Souza et al. 2008; Siqueira et al. 2010). In human hepatoma cells, MMP9 expression is induced by Sdc-4/HS-dependent signaling via SDF-1 and its receptor CXCR4 (Sutton et al. 2007). In contrast, glypican-3 expression protected from upregulation of MMP expression in a syngeneic mouse model of breast cancer (Peters et al. 2003). Our group could recently show that overexpression of a constitutively membrane-bound form of Sdc-1 increased secreted amounts of the endogenous MMP inhibitor TIMP-1, whereas overexpression of soluble Sdc-1 reduced TIMP-1 expression, resulting in increased invasiveness of human MCF-7 breast cancer cells (Nikolova et al. 2009). MMP-mediated shedding of Sdc-1 may therefore mark a switch from a proliferative to an invasive phenotype in breast cancer cells. Apart from regulatory aspects, soluble Sdc-1 modulates the activity of proteases and antiproteases via direct HS-dependent binding interactions in the context of wound repair and angiogenesis (Bernfield et al. 1999; Elenius et al. 2004). Moreover, a HS-dependent inhibitory function of Sdc-2 on MMP2 activity has been demonstrated in Lewis lung carcinoma cells (Munesue et al. 2007). In addition to MMPs, increased activity and expression of heparanase, a β-endoglucuronidase, which cleaves glycosidic bonds in heparan sulfate, promotes metastasis in vitro and in vivo (Götte and Yip 2006; Casu et al. 2008; Fux et al. 2009). Heparanase expression correlates with tumor metastasis in clinical specimens (Maxhimer et al. 2002; Lerner et al. 2008). Degradation of HS by heparanase promotes metastasis via several molecular mechanisms, including physical removal of sterical hindrances upon basement membrane HSPG degradation, release of biologically active HS fragments capable of activating growth factor signaling, release of cytokines and angiogenic factors, and direct induction of signaling, possibly even via regulation of transcriptional activity in the nucleus (Götte and Yip 2006; Fux et al. 2009; Zong et al. 2009). Metastasis also depends on the dissemination of invasive tumor cells into the circulation and encompasses adhesive interactions with endothelial cells, leukocytes, and platelets, ultimately resulting in the colonization of distant tissues and organs by the tumor cells (cf. Chaps. 1 and 3). We have recently shown that a loss of Sdc-1, an important modulator of chemokine function (Götte 2003), results in increased leukocyte–endothelial interactions both in vitro and in vivo, due to a HS-dependent modulation of integrin–ICAM-1 interactions (Kharabi Masouleh et al. 2009; Floer et al. 2010). Although still speculative at this point, Sdc-1 may regulate adhesion of tumor cells to blood and lymphatic vessel endothelium or promote the association with different host cells during metastatic seeding. Finally, the interplay of tumor cell-derived Sdc-1 and heparanase can even result in a remodeling of the target organ of metastasing tumor cells. This was recently shown by Kelly et al. (2010), who demonstrated that heparanase-mediated shedding and processing of the HS chains of Sdc-1 in MCF-7 breast cancer cells result in an IL-8-dependent stimulation of osteoclastogenesis and bone destruction in vivo.

6.2.3 HSPGs and Tumor Angiogenesis

In order to grow beyond a diameter of 2 mm^2, primary tumors and metastases depend on nutrient support from the vascular system (Folkman 2007). As outlined in Sect. 1.3, HSPGs regulate the process of angiogenesis and the formation of blood vessels from existing vessels, via interaction with angiogenic factors such as vascular endothelial growth factors (VEGFs), platelet-derived growth factor, FGFs, angiopoietins, endothelins, and their receptors (Götte 2003; Iozzo et al. 2009). Apart from structural modifications in NDST1-deficient mice (Sect. 1.3), alterations in 6-O-sulfation of HS have an impact on tumor angiogenesis, as xenografts derived from HSulf-1-expressing breast cancer cells showed reduced angiogenesis in vivo, due to reduced signaling mediated by the HS-dependent angiogenic factors FGF-2, HGF, and heparin-binding VEGF isoforms (Narita et al. 2006). A proangiogenic role has also been documented for heparanase, which can convert HS into highly angiogenic fragments (Roy and Marchetti 2009; Kato et al. 1998). As one of several alternative angiogenic mechanisms, heparanase can induce shedding of Sdc-1 in a variety of tumor cell lines, generating soluble Sdc-1–VEGF complexes that promote tumor angiogenesis (Purushothaman et al. 2010). Sdc-1 expression has been correlated to tumor angiogenesis in clinical samples and xenograft studies (Götte et al. 2007; Maeda et al. 2006; Purushothaman et al. 2010), and Sdc-1 has been proposed to regulate angiogenesis via HS-dependent recruitment of leukocytes secreting angiogenic factors (Götte 2003; Kharabi Masouleh et al. 2009; Floer et al. 2010), and by controlling the proteolytic milieu during tissue remodeling (Elenius et al. 2004). Moreover, the core protein of Sdc-1 regulates avβ3 and avβ5 integrin activation during tumor angiogenesis in vivo, with implications for Sdc-1-peptide inhibitor-based antiangiogenic therapy (Purushothaman et al. 2010; Beauvais et al. 2009). The angiogenesis-modulating function of Sdc-1, which correlates with reduced tumor-derived VEGF levels and angiogenesis, appears to be one mechanistic aspect of reduced multifocal metastasis of Sdc-1 depleted myeloma cells in vivo (Khotskaya et al. 2009). Sdc-2 appears to modulate angiogenesis via related molecular mechanisms (Noguer and Reina 2009). In addition to cell surface HSPGs, antisense inhibition of the basement membrane

PG perlecan led to decreased colon and prostate carcinoma growth and tumor angiogenesis (Iozzo et al. 2009; Savorè et al. 2005). Impaired angiogenesis and retarded tumor growth were also seen in perlecan-deficient mice (Iozzo et al. 2009). Both HS-dependent effects on signal transduction via HS-binding angiogenic factors such as VEGF and angiopoietin isoforms and an integrin-targeting angiogenesis-inhibiting function of core-protein fragments have been discussed as molecular mechanisms (Xu et al. 2004; Savorè et al. 2005; Bix et al. 2006). As the HGF antagonist NK4 inhibits both HGF-dependent and perlecan-based angiogenic mechanisms, it may represent a particularly efficient potential therapeutic agent in antiangiogenic cancer therapy (Suzuki et al. 2010).

6.3 Heparan Sulfate-Centered Approaches of Cancer Therapy

The involvement of HSPGs and their biosynthetic and processing enzymes in malignant transformation, tumor angiogenesis, and metastasis has promoted the development of drugs targeting the malignancy-associated biological functions of these molecules. Examples for different therapeutic strategies are presented in the following paragraphs.

6.3.1 Heparin and Heparinoids

As outlined in Chap. 4, heparin is a potent anticoagulant that has been used to treat and prevent thromboembolism for decades. The anticoagulant activity appears to affect cancer progression by decreasing thrombin generation and fibrin formation (Prandoni et al. 2005). In addition, the inhibitory action of heparin on heparanase activity (see below), integrin-mediated tumor cell adhesion and motility, competitive binding to cytokines and chemokines, as well as interference with P- and L-selectin functions (cf Chap. 1) contribute to its beneficial effect in cancer therapy (Götte 2003; Prandoni et al. 2005; Borsig 2010). For example, antimetastatic properties of different heparin preparations were demonstrated in animal models of prostate cancer, colon carcinoma, breast cancer, lung cancer, osteosarcoma, and melanoma (Borsig 2010). Heparin proved to be most efficient when it was applied around the time of intravenous tumor cell injection and when the amount of heparin applied was at least at the clinically used concentrations (Borsig 2010). Although the anticoagulant properties of heparin may mechanistically contribute to some of its antitumoral activities, anticoagulation constitutes a potential adverse side effect in cancer therapy. Therefore, the identification and characterization of heparin derivatives with low anticoagulant properties and high antitumoral activity has gained a lot of attention. In the syngeneic B16F10 melanoma mouse model of metastasis, a non-anticoagulant heparin derivative inhibited the number of metastases by 58% (Kragh et al. 2005), whereas a low anticoagulant heparin was as effective as conventional heparin in the inhibition of lung colonization by tumor cells in a different syngeneic model (Angelini et al. 2005). A modulation of circulating tumor cell adhesion to the vasculature may be part of the molecular mechanism of heparin action (Borsig 2010). For example, knockout mouse studies demonstrated that melanoma cells directly interact with postcapillary venules in a P-selectin-dependent fashion and that heparin inhibits P-selectin-mediated melanoma cell rolling and metastasis formation (Ludwig et al. 2004). Of note, the use of a non-anticoagulant N-acetylated, glycol-split heparin effectively blocked myeloma growth in vivo (Casu et al. 2008).

While most in vivo studies demonstrate an antimetastatic effect, but rather little activity against the primary tumor, the combined use of heparin and conventional chemotherapy may hold some promise. Unfractionated heparin potentiated doxorubicin accumulation and cytotoxicity in a human breast cancer cell line in vitro, due to inhibition of multidrug resistance mechanisms (Angelini et al. 2005), and combination of heparin with a 5-fluorouracil, mitomycin C, and cisplatin reduced liver metastasis, following curative resection of pancreatic cancer (Aiura et al. 2010). Moreover, treating small cell lung cancer patients with a combination of deltaparin, cyclophosphamide, epirubicine, and vincristine led to a significant improvement of overall tumor response rates, median progression-free survival, and median overall survival compared with chemotherapy alone (Altinbas et al. 2004). Several clinical studies, including phase III trials, have been performed to assess the potentially beneficial effect of heparin on cancer patient survival (Gerotziafas et al. 2008; Kuderer et al. 2007; Griffiths et al. 2009). While most of these studies demonstrate beneficial effects with respect to overall survival, some histological types of cancers appear to benefit more from heparin treatment than others: Gerotziafas et al. (2008) concluded that heparin treatment in small cell lung cancer or pancreatic cancer has probably a more clear beneficial effect than in others, e.g., colorectal or breast cancer, and that survival may be improved to a larger extent in patients with longer life expectancy compared to those who have shorter life expectancy. Large, carefully designed phase III trials such as the FRAGMATIC study (Griffiths et al. 2009), which aims at recruiting 2,200 patients, will help to fully assess the clinical benefit of heparin treatment in cancer.

6.3.2 Inhibitors of Heparan Sulfate Biosynthesis

Due to the importance of HS for a multitude of physiological processes, systemic inhibition of HS biosynthesis may very likely result in undesired side effects in a clinical setting. Nevertheless, the antimetastatic reagent 5-hexyl-2-deoxyuridine, which reduces biosynthesis of HS and other glycoconjugates by inhibiting conversion of glucosamine to UDP-sugars (Jeney et al. 1990), decreased formation of metastatic liver colonies in vivo (Tímár et al. 1995). Moreover, Belting et al. (2002) could demonstrate that combinatorial application of difluoromethylornithine-mediated polyamine depletion and xyloside inhibition of glycosaminoglycan synthesis resulted in reduced tumor growth in vivo. While these studies in animal models are promising, potential side effects need to be carefully evaluated before these studies can be translated into clinical applications.

6.3.3 Peptide-Based Inhibition of HSPG Function

In contrast to the more broad effects of a heparin-based therapy, peptide-based competitive inhibition approaches allow for a more selective targeting of unique HS-binding ligands, which may be associated with a more beneficial side-effect profile in clinical applications. For example, thrombospondin-1-derived heparin-binding peptides are capable of inducing promyelocytic leukemia cell apoptosis (Bruel et al. 2005). Moreover, a fusion protein composed of the heparin-binding domain of neuregulin and a soluble form of the EGFR Her4 blocked autocrine and paracrine signaling and reduced the proliferation of MCF10CA1 breast cancer cells (Ma et al. 2009). While not targeting the glycan moiety, but rather the integrin-activating domain of the core protein of the HSPG Sdc-1, the peptide drug synstatin has demonstrated its inhibitory potential on myeloma and breast cancer cell growth and tumor angiogenesis in animal models (Purushothaman et al. 2010; Beauvais et al. 2009) Therefore, synstatin may be a prime candidate for combinatorial targeting of HS-dependent and HS-independent functions of Sdc-1 in cancer.

6.3.4 Inhibitors of Heparan Sulfate-Processing Enzymes

Several of the tumorigenesis-related functions of HSPGs depend on the interplay with degradative enzymes, including heparanase, HSulf-1, and HSulf-2. Heparanase has emerged as a major therapeutic target in this context (Fux et al. 2009; Casu et al. 2008; Cohen et al. 2006). Studies in the RIP-Tag2 mouse model of pancreatic cancer could show that PI-88/muparfostat (phosphomannopentose sulfate), a sulfated oligosaccharide mimetic of heparan sulfate that cannot be cleaved by heparanase, reduced early progenitor lesions and inhibited tumor growth at late stages (Parish et al. 1999), targeting tumor cell proliferation and angiogenesis and stimulating apoptosis. PI-88 also inhibited rat adenocarcinoma cell growth and metastasis to lymph nodes in an in vivo model (Parish et al. 1999). Of note, PI-88 does not only inhibit heparanase, but also HSulf-1 and HSulf-2 activity (Hossain et al. 2010). Currently, PI-88 is evaluated in several phase I–III clinical trials on advanced melanoma, hepatocellular carcinoma, nonsmall cell lung cancer, and prostate cancer (Chow et al. 2008; Liu et al. 2009; Lewis et al. 2008). According to the results of a randomized phase II trial, PI-88 was safe and showed preliminary efficacy as an adjunct therapy for postoperative hepatocellular carcinoma (Liu et al. 2009). Furthermore, some evidence of activity was reported in a phase II study in patients with advanced melanoma (Lewis et al. 2008); however, a multicenter phase I/II study of PI-88 in combination with docetaxel in patients with metastatic castrate-resistant prostate cancer had to be terminated due to significant hematologic toxicity (Khasraw et al. 2010). It remains to be shown if lipid-modified polysulfated oligosaccharide glycosides with reduced anticoagulant activity show a more favorable toxicity profile in clinical studies (Johnstone et al. 2010).

As previously mentioned, heparin is an inhibitor of heparanase activity. Naggi et al. (2005) demonstrated that N-desulfation and N-acetylation, as well as glycol-splitting of nonsulfated uronic acids of heparin, increased its inhibitory activity on heparanase. These inhibitors have recently been successfully applied in an in vivo model of myeloma (Casu et al. 2008). In addition, laminarin sulfate, suramin, and its analogues, as well as different low-molecular-weight heparins, exhibit inhibitory activity on heparanase and have demonstrated anticancer activity in different in vitro and in vivo studies (Yip et al. 2006; Götte and Yip 2006; Casu et al. 2008; Borsig 2010).

6.3.5 HS and HSPGs as Angiogenesis Inhibitors

The pharmacological inhibition of angiogenesis is a well-established concept in cancer therapy. Antiangiogenic therapy has a favorable toxicity profile compared to conventional chemotherapy, since physiological angiogenesis in adults is restricted to a limited number of specialized processes, such as wound repair and the menstrual cycle (cf Chap. 1). Moreover, antiangiogenesis is associated with a lower risk of developing drug resistance due to its indirect mode of action on tumor cell growth. Different molecular approaches have been taken to target the function of HSPGs and heparanase in tumor angiogenesis (cf Sect. 1.1.3). The first approach relies on a competitive inhibition of the interaction of HS with angiogenic factors by heparin and heparinoids: For example, tinzaparin, low-molecular-weight N,O-sulfated *E. coli* K5 polysaccharide derivatives, and glycol-split heparins inhibited antiangiogenic activity in endothelial cell sprouting and CAM angiogenesis assays (reviewed in Yip et al. 2006), which was mechanistically linked to a disruption of angiogenic cytokine signaling through their respective receptor tyrosine kinases. An alternative approach is the inhibition of the angiogenic activity of heparanase (cf Sect. 6.3.4). In some cases, this approach overlaps with the inhibition of angiogenic signaling, as the heparanase inhibitor PI-88 also interferes with the bioavailability and function of HS-binding angiogenic factors (Parish et al. 1999). In vivo studies revealed a successful inhibition of tumor angiogenesis in highly invasive rat mammary adenocarcinomas, in a rodent model of human myeloid leukemia, and in the RIP-Tag2 model of multistage pancreatic islet carcinogenesis by PI-88 (Iversen et al. 2002; Joyce et al. 2005). Most recently, interference with Sdc-1-mediated integrin activation by the peptide synstatin has emerged as an additional HSPG-centered antiangiogenic approach, which has proved to be successful in animal models of myeloma and breast cancer (Beauvais et al. 2009; Purushothaman et al. 2010).

6.4 Perspective

Heparan sulfate and HSPG modulate all relevant stages of tumor progression (Fig. 6.1). The frequent dysregulation of HSPGs and of HS-processing enzymes in malignant disease and the establishment of prognostic values for several HSPGs underscore the oncological relevance of these glycoproteins from both translational and clinical perspectives. However, the complexity of physiological HSPG functions is reflected by the diversity of their roles in tumor progression. Depending on a variety of parameters, including HS fine structure, the nature and location of the core protein (membrane-bound versus soluble), the cell type expressing the HSPG (epithelial versus stromal), and the absence or presence of heparan sulfate-processing enzymes, HSPGs play different, and sometimes opposing, roles in tumor initiation, progression, and metastasis. Moreover, a particular HSPG may have a different function at different stages of tumor progression. For example, tumor cells need to loosen cell–cell and cell–matrix contacts in order to escape from the primary tumor mass and to become metastatic, but they have to establish contact to the vascular endothelium in order to escape from the circulation during the metastatic process. The transmembrane form of a HSPG may act as a coreceptor for mitogenic signaling, while its soluble, shed form acts as a competitive inhibitor. Changes in HS fine structure, in HSPG core protein expression, or conversion of membrane-bound HSPG into soluble ectodomains may aid the tumor cell in adapting to different stages of the pathophysiological process. From a therapeutic perspective, this functional diversity constitutes a major challenge. Therefore, future investigations need to carefully assess the role of defined HSPGs, HS structures, and HS-processing enzymes in molecular detail according to tumor type, relevant cell types (such as tumor, stroma, endothelium or leukocytes), and the stage of tumor progression as a solid foundation for HS-centered therapies. Novel diagnostic tools such as glycoprofiling array techniques and increased sensitivity of analytical techniques including mass spectrometry will increase the knowledge on the detailed role of specific HS structures in cancer, aiding the design of next-generation glycotherapeutics. A combination of conventional pharmaceutics with glycotherapeutics may achieve the greatest therapeutic effect. The encouraging results of in vivo studies in animal models as well as clinical trials demonstrate the potential of HS-targeted based drugs and their future use in patient management.

Acknowledgments Work in the author's laboratory on the role of heparan sulfate proteoglycans in cancer is financially supported by grants of the German Academic Exchange Service DAAD A/06/90277 (SAI), the Deutsche Forschungsgemeinschaft/CNPq 444 BRA-113/63/0-1, Münster University Hospital, Innovative Medizinische Forschung' IMF GÖ 320901, and the Maria Möller Stiftung (MG).

References

Abiatari I, Kleeff J, Li J, Felix K, Büchler MW, Friess H (2006) Hsulf-1 regulates growth and invasion of pancreatic cancer cells. J Clin Pathol 59:1052–1058

Aiura K, Takahashi S, Matsui J, Ueda M, Kitagawa Y (2010) Beneficial effects of 5-Fluorouracil and heparin-based portal infusion chemotherapy combined with Mitomycin C and Cisplatin after curative resection of pancreatic cancer. Pancreatology 10:250–258

Altinbas M, Coskun H, Er O, Ozkan M, Eser B, Unal A, Cetin M, Soyuer S (2004) A randomized clinical trial of combination chemotherapy with and without low-molecular-weight heparin in small cell lung cancer. J Thromb Haemost 2:1266–1271

Angelini A, Di Febbo C, Ciofani G, Di Nisio M, Baccante G, Di Ilio C, Cuccurullo F, Porreca E (2005) Inhibition of P-glycoprotein-mediated multidrug resistance by unfractionated heparin: a new potential chemosensitizer for cancer therapy. Cancer Biol Ther 4:313–317

Beauvais DM, Burbach B, Rapraeger AC (2004) The syndecan-1 ectodomain regulates alphavbeta3 integrin activity in human mammary carcinoma cells. J Cell Biol 167:171–181

Beauvais DM, Ell BJ, McWhorter AR, Rapraeger AC (2009) Syndecan-1 regulates alphavbeta3 and alphavbeta5 integrin activation during angiogenesis and is blocked by synstatin, a novel peptide inhibitor. J Exp Med 206:691–705

Belting M, Borsig L, Fuster MM, Brown JR, Persson L, Fransson LA, Esko JD (2002) Tumor attenuation by combined heparan sulfate and polyamine depletion. Proc Natl Acad Sci USA 99:371–376

Bernfield M, Götte M, Park PW, Reizes O, Fitzgerald ML, Lincecum J, Zako M (1999) Functions of cell surface heparan sulfate proteoglycans. Annu Rev Biochem 68:729–777

Bix G, Castello R, Burrows M, Zoeller JJ, Weech M, Iozzo RA, Cardi C, Thakur ML, Barker CA, Camphausen K, Iozzo RV (2006) Endorepellin in vivo: targeting the tumor vasculature and retarding cancer growth and metabolism. J Natl Cancer Inst 98:1634–1646

Borsig L (2010) Antimetastatic activities of heparins and modified heparins. Experimental evidence. Thromb Res 125(Suppl 2):S66–S71

Bruel A, Touhami-Carrier M, Thomaidis A, Legrand C (2005) Thrombospondin-1 (TSP-1) and TSP-1-derived heparin-binding peptides induce promyelocytic leukemia cell differentiation and apoptosis. Anticancer Res 25:757–764

Bui C, Ouzzine M, Talhaoui I, Sharp S, Prydz K, Coughtrie MW, Fournel-Gigleux S (2010) Epigenetics: methylation-associated repression of heparan sulfate 3-O-sulfotransferase gene expression contributes to the invasive phenotype of H-EMC-SS chondrosarcoma cells. FASEB J 24:436–450

Casu B, Vlodavsky I, Sanderson RD (2008) Non-anticoagulant heparins and inhibition of cancer. Pathophysiol Haemost Thromb 36:195–203

Chow LQ, Gustafson D, O'Bryant CL, Gore L, Basche M, Holden SN, Morrow MC, Grolnic S, Creese BR, Roberts KL, Davis K, Addison R, Eckhardt SG (2008) A phase I pharmacological and biological study of PI-88 and docetaxel in patients with advanced malignancies. Cancer Chemother Pharmacol 63:65–74

Cohen I, Pappo O, Elkin M, San T, Bar-Shavit R, Hazan R, Peretz T, Vlodavsky I, Abramovitch R (2006) Heparanase promotes growth, angiogenesis and survival of primary breast tumors. Int J Cancer 118:1609–1617

Contreras HR, Ledezma R, Vergara J, Cifuentes F, Barra C, Cabello P, Gallegos I, Morales B, Huidobro C, Castellón EA (2010) The expression of syndecan-1 and -2 is associated with Gleason score and epithelial-mesenchymal transition markers, E-cadherin and beta-catenin, in prostate cancer. Urol Oncol 28:534–540

Derksen PW, K R, Evers LM, van Oers MH, Spaargaren M, Pals ST (2002) Cell surface proteoglycan syndecan-1 mediates hepatocyte growth factor binding and promotes Met signaling in multiple myeloma. Blood 99:1405–1410

Dhodapkar MV, Abe E, Theus A, Lacy M, Langford JK, Barlogie B, Sanderson RD (1998) Syndecan-1 is a multifunctional regulator of myeloma pathobiology: control of tumor cell survival, growth, and bone cell differentiation. Blood 91:2679–2688

Dieudonné FX, Marion A, Haÿ E, Marie PJ, Modrowski D (2010) High Wnt signaling represses the proapoptotic proteoglycan syndecan-2 in Osteosarcoma cells. Cancer Res 70:5399
Egeblad M, Werb Z (2002) New functions for the matrix metalloproteinases in cancer progression. Nat Rev Cancer 2:161–174
Elenius V, Götte M, Reizes O, Elenius K, Bernfield M (2004) Inhibition by the soluble syndecan-1 ectodomains delays wound repair in mice overexpressing syndecan-1. J Biol Chem 279: 41928–41935
Fernig DG, Chen H, Rahmoune H, Descamps S, Boilly B, Hondermarck H (2000) Differential regulation of FGF-1 and -2 mitogenic activity is related to their kinetics of binding to heparan sulfate in MDA-MB-231 human breast cancer cells. Biochem Biophys Res Commun 267:770–776
Floer M, Götte M, Wild MK, Heidemann J, Gassar ES, Domschke W, Kiesel L, Luegering A, Kucharzik T (2010) Enoxaparin improves the course of dextran sodium sulfate-induced colitis in syndecan-1-deficient mice. Am J Pathol 176:146–157
Folkman J (2007) Angiogenesis: an organizing principle for drug discovery? Nat Rev Drug Discov 6:273–286
Franzen CA, Chen C, Todorović V, Juric V, Monzon RI, Lau LF (2009) Matrix protein CCN1 is critical for prostate carcinoma cell proliferation and TRAIL-induced apoptosis. Mol Cancer Res 7:1045–1055
Fulda S (2009) Tumor resistance to apoptosis. Int J Cancer 124:511–515
Fuster MM, Esko JD (2005) The sweet and sour of cancer: glycans as novel therapeutic targets. Nat Rev Cancer 5:526–542
Fux L, Ilan N, Sanderson RD, Vlodavsky I (2009) Heparanase: busy at the cell surface. Trends Biochem Sci 34:511–519
Gama-de-Souza LN, Cyreno-Oliveira E, Freitas VM, Melo ES, Vilas-Boas VF, Moriscot AS, Jaeger RG (2008) Adhesion and protease activity in cell lines from human salivary gland tumors are regulated by the laminin-derived peptide AG73, syndecan-1 and beta1 integrin. Matrix Biol 27:402–419
Gerotziafas GT, Papageorgiou C, Hatmi M, Samama MM, Elalamy I (2008) Clinical studies with anticoagulants to improve survival in cancer patients. Pathophysiol Haemost Thromb 36: 204–211
Gonzalez AD, Kaya M, Shi W, Song H, Testa JR, Penn LZ, Filmus J (1998) OCI-5/GPC3, a glypican encoded by a gene that is mutated in the Simpson-Golabi-Behmel overgrowth syndrome, induces apoptosis in a cell line-specific manner. J Cell Biol 141:1407–1414
Götte M (2003) Syndecans in inflammation. FASEB J 17:575–591
Götte M, Yip G (2006) Heparanase, hyaluronan, and CD44 in cancers: a breast carcinoma perspective. Cancer Res 66:10233–10237
Götte M, Kersting C, Radke I, Kiesel L, Wülfing P (2007) An expression signature of syndecan-1 (CD138), E-cadherin and c-met is associated with factors of angiogenesis and lymphangiogenesis in ductal breast carcinoma in situ. Breast Cancer Res 9:R8
Götte M, Ibrahim S, Yip G, Stock CM, Kiesel L (2010) Syndecan-1 (CD138) modulates proliferation, motility and invasiveness of breast cancer cells via HGF/c-Met-dependent and independent signaling pathways. Onkologie 33(suppl 2):10-10
Griffiths GO, Burns S, Noble SI, Macbeth FR, Cohen D, Maughan TS (2009) FRAGMATIC: a randomised phase III clinical trial investigating the effect of fragmin added to standard therapy in patients with lung cancer. BMC Cancer 9:355
Hasengaowa KJ, Kusumoto T, Shinyo Y, Seki N, Nakamura K, Hongo A, Hiramatsu Y (2005) Loss of basement membrane heparan sulfate expression is associated with tumor progression in endometrial cancer. Eur J Gynaecol Oncol 26:403–406
Hossain MM, Hosono-Fukao T, Tang R, Sugaya N, van Kuppevelt TH, Jenniskens GJ, Kimata K, Rosen SD, Uchimura K (2010) Direct detection of HSulf-1 and HSulf-2 activities on extracellular heparan sulfate and their inhibition by PI-88. Glycobiology 20:175–186

Ikeguchi M, Hirooka Y, Kaibara N (2003) Heparanase gene expression and its correlation with spontaneous apoptosis in hepatocytes of cirrhotic liver and carcinoma. Eur J Cancer 39: 86–90

Iozzo RV, Zoeller J, Nyström A (2009) Basement membrane proteoglycans: modulators Par Excellence of cancer growth and angiogenesis. Mol Cells 27:503–513

Ishikawa T, Kramer RH (2010) Sdc1 negatively modulates carcinoma cell motility and invasion. Exp Cell Res 316:951–965

Iversen PO, Sorenson D, Benestad HB (2002) Inhibitors of angiogenesis selectively reduce the malignant cell load in rodent models of human myeloid leukemias. Leukemia 16:376–381

Jayson GC, Lyon M, Paraskeva C, Turnbull JE, Deakin JA, Gallagher JT (1998) Heparan sulfate undergoes specific structural changes during the progression from human colon adenoma to carcinoma in vitro. J Biol Chem 273:51–57

Jeney A, Timar J, Pogány G, Paku S, Moczár E, Mareel M, Otvös L, Kopper L, Lapis K (1990) Glycosaminoglycans as novel target in antitumor therapy. Tokai J Exp Clin Med 15:167–177

Johnstone KD, Karoli T, Liu L, Dredge K, Copeman E, Li CP, Davis K, Hammond E, Bytheway I, Kostewicz E, Chiu FC, Shackleford DM, Charman SA, Charman WN, Harenberg J, Gonda TJ, Ferro V (2010) Synthesis and biological evaluation of polysulfated oligosaccharide glycosides as inhibitors of angiogenesis and tumor growth. J Med Chem 53:1686–1699

Joyce JA, Freeman C, MeyerMorse N, Parish CR, Hanahan D (2005) A functional heparan sulfate mimetic implicates both heparanase and heparan sulfate in tumor angiogenesis and invasion in a mouse model of multistage cancer. Oncogene 24:4037–4051

Kamimura K, Rhodes J, Ueda R, McNeely M, Shukla D, Kimata K, Spear PG, Shworak NW, Nakato H (2004) Regulation of Notch signaling by Drosophila heparan sulfate 3-O sulfotransferase. J Cell Biol 166:1069–1079

Kato M, Wang H, Kainulainen V, Fitzgerald ML, Ledbetter S, Ornitz DM, Bernfield M (1998) Physiological degradation converts the soluble syndecan-1 ectodomain from an inhibitor to a potent activator of FGF-2. Nat Med 4:691–697

Kelly T, Suva L, Nicks KM, MacLeod V, Sanderson RD (2010) Tumor-derived syndecan-1 mediates distal cross-talk with bone that enhances osteoclastogenesis. J Bone Miner Res 25:1295–1304

Kharabi Masouleh B, Ten Dam GB, Wild MK, Seelige R, van der Vlag J, Rops AL, Echtermeyer FG, Vestweber D, van Kuppevelt TH, Kiesel L, Götte M (2009) Role of the heparan sulfate proteoglycan syndecan-1 (CD138) in delayed-type hypersensitivity. J Immunol 182: 4985–4993

Khasraw M, Pavlakis N, McCowatt S, Underhill C, Begbie S, de Souza P, Boyce A, Parnis F, Lim V, Harvie R, Marx G (2010) Multicentre phase I/II study of PI-88, a heparanase inhibitor in combination with docetaxel in patients with metastatic castrate-resistant prostate cancer. Ann Oncol 21:1302–1307

Khotskaya YB, Dai Y, Ritchie JP, MacLeod V, Yang Y, Zinn K, Sanderson RD (2009) Syndecan-1 is required for robust growth, vascularization, and metastasis of myeloma tumors in vivo. J Biol Chem 284:26085–26095

Kim H, Xu G, Borczuk AC, Busch S, Filmus J, Capurro M, Brody JS, Lange J, D'Armiento JM, Rothman PB, Powell CA (2003) The heparan sulfate proteoglycan GPC3 is a potential lung tumor suppressor. Am J Respir Cell Mol Biol 29:694–701

Kodama J, Shinyo Y, Hasengaowa KT, Seki N, Nakamura K, Hongo A, Hiramatsu Y (2005) Loss of basement membrane heparan sulfate expression is associated with pelvic lymph node metastasis in invasive cervical cancer. Oncol Rep 14:89–92

Kragh M, Binderup L, Vig Hjarnaa PJ, Bramm E, Johansen KB, Frimundt Petersen C (2005) Non-anti-coagulant heparin inhibits metastasis but not primary tumor growth. Oncol Rep 14:99–104

Kuderer NM, Khorana A, Lyman GH, Francis CW (2007) A meta-analysis and systematic review of the efficacy and safety of anticoagulants as cancer treatment: impact on survival and bleeding complications. Cancer Biol Ther 110:1149–1161

Kusche-Gullberg M, Kjellen L (2003) Sulfotransferases in glycosaminoglycan biosynthesis. Curr Opin Struct Biol 13:605–611

Lai J, Chien J, Staub J, Avula R, Greene EL, Matthews TA, Smith DI, Kaufmann SH, Roberts LR, Shridhar V (2003) Loss of HSulf-1 up-regulates heparin-binding growth factor signaling in cancer. J Biol Chem 278:23107–23117

Lai JP, Chien J, Moser DR, Staub JK, Aderca I, Montoya DP, Matthews TA, Nagorney DM, Cunningham JM, Smith DI, Greene EL, Shridhar V, Roberts LR (2004a) hSulf1 Sulfatase promotes apoptosis of hepatocellular cancer cells by decreasing heparin-binding growth factor signaling. Gastroenterology 126:231–248

Lai JP, Chien J, Strome SE, Staub J, Montoya DP, Greene EL, Smith DI, Roberts LR, Shridhar V (2004b) HSulf-1 modulates HGF-mediated tumor cell invasion and signaling in head and neck squamous carcinoma. Oncogene 23:1439–1447

Lai JP, Yu C, Moser CD, Aderca I, Han T, Garvey TD, Murphy LM, Garrity-Park MM, Shridhar V, Adjei AA, Roberts LR (2006) SULF1 inhibits tumor growth and potentiates the effects of histone deacetylase inhibitors in hepatocellular carcinoma. Gastroenterology 130:2130–2144

Lemjabbar-Alaoui H, van Zante A, Singer MS, Xue Q, Wang YQ, Tsay D, He B, Jablons DM, Rosen SD (2010) Sulf-2, a heparan sulfate endosulfatase, promotes human lung carcinogenesis. Oncogene 29:635–646

Lerner I, Baraz L, Pikarsky E, Meirovitz A, Edovitsky E, Peretz T, Vlodavsky I, Elkin M (2008) Function of heparanase in prostate tumorigenesis: potential for therapy. Clin Cancer Res 14:668–676

Lewis KD, Robinson W, Millward MJ, Powell A, Price TJ, Thomson DB, Walpole ET, Haydon AM, Creese BR, Roberts KL, Zalcberg JR, Gonzalez R (2008) A phase II study of the heparanase inhibitor PI-88 in patients with advanced melanoma. Invest New Drugs 26:89–94

Liu BY, Kim Y, Leatherberry V, Cowin P, Alexander CM (2003) Mammary gland development requires syndecan-1 to create a beta-catenin/TCF-responsive mammary epithelial subpopulation. Oncogene 22:9243–9253

Liu CJ, Lee P, Lin DY, Wu CC, Jeng LB, Lin PW, Mok KT, Lee WC, Yeh HZ, Ho MC, Yang SS, Lee CC, Yu MC, Hu RH, Peng CY, Lai KL, Chang SS, Chen PJ (2009) Heparanase inhibitor PI-88 as adjuvant therapy for hepatocellular carcinoma after curative resection: a randomized phase II trial for safety and optimal dosage. J Hepatol 50:958–968

Ludwig RJ, Boehme B, Podda M, Henschler R, Jager E, Tandi C, Boehncke WH, Zollner TM, Kaufmann R, Gille J (2004) Endothelial P-selectin as a target of heparin action in experimental melanoma lung metastasis. Cancer Res 64:2743–2750

Luyten A, Mortier E, Van Campenhout C, Taelman V, Degeest G, Wuytens G, Lambaerts K, David G, Bellefroid EJ, Zimmermann P (2008) The postsynaptic density 95/disc-large/zona occludens protein syntenin directly interacts with frizzled 7 and supports noncanonical Wnt signaling. Mol Biol Cell 19:1594–1604

Ma Z, Li Q, An H, Pankonin MS, Wang J, Loeb JA (2009) Targeting human epidermal growth factor receptor signaling with the neuregulin's heparin-binding domain. J Biol Chem 284: 32108–32115

Maeda T, Alexander C, Friedl A (2004) Induction of syndecan-1 expression in stromal fibroblasts promotes proliferation of human breast cancer cells. Cancer Res 64:612–621

Maeda T, Desouky J, Friedl A (2006) Syndecan-1 expression by stromal fibroblasts promotes breast carcinoma growth in vivo and stimulates tumor angiogenesis. Oncogene 25:1408–1412

Mahtouk K, Cremer F, Rème T, Jourdan M, Baudard M, Moreaux J, Requirand G, Fiol G, De Vos J, Moos M, Quittet P, Goldschmidt H, Rossi JF, Hose D, Klein B (2006) Heparan sulphate proteoglycans are essential for the myeloma cell growth activity of EGF-family ligands in multiple myeloma. Oncogene 25:7180–7191

Matsuda K, Maruyama H, Guo F, Kleeff J, Itakura J, Matsumoto Y, Lander AD, Korc M (2001) Glypican-1 is overexpressed in human breast cancer and modulates the mitogenic effects of multiple heparin-binding growth factors in breast cancer cells. Cancer Res 61:5562–5569

Maxhimer JB, Quiros R, Stewart R, Dowlatshahi K, Gattuso P, Fan M, Prinz RA, Xu X (2002) Heparanase-1 expression is associated with the metastatic potential of breast cancer. Surgery 132:326–333

McDermott SP, Ranheim E, Leatherberry VS, Khwaja SS, Klos KS, Alexander CM (2007) Juvenile syndecan-1 null mice are protected from carcinogen-induced tumor development. Oncogene 26:1407–1416

Micalizzi DS, Farabaugh S, Ford HL (2010) Epithelial-mesenchymal transition in cancer: parallels between normal development and tumor progression. J Mammary Gland Biol Neoplasia 15 (2):117–134

Modrowski D, Orosco A, Thévenard J, Fromigué O, Marie PJ (2005) Syndecan-2 overexpression induces osteosarcoma cell apoptosis: implication of syndecan-2 cytoplasmic domain and JNK signaling. Bone 37:180–189

Morgan MR, Humphries M, Bass MD (2007) Synergistic control of cell adhesion by integrins and syndecans. Nat Rev Mol Cell Biol 8:957–969

Munesue S, Yoshitomi Y, Kusano Y, Koyama Y, Nishiyama A, Nakanishi H, Miyazaki K, Ishimaru T, Miyaura S, Okayama M, Oguri K (2007) A novel function of syndecan-2, suppression of matrix metalloproteinase-2 activation, which causes suppression of metastasis. J Biol Chem 282:28164–28174

Naggi A, Casu B, Perez M, Torri G, Cassinelli G, Penco S, Pisano C, Giannini G, Ishai-Michaeli R, Vlodavsky I (2005) Modulation of the heparanase-inhibiting activity of heparin through selective desulfation, graded N-acetylation, and glycol splitting. J Biol Chem 280: 12103–12113

Narita K, Staub J, Chien J, Meyer K, Bauer M, Friedl A, Ramakrishnan S, Shridhar V (2006) HSulf-1 inhibits angiogenesis and tumorigenesis in vivo. Cancer Res 66:6025–6032

Nikolova V, Koo C, Ibrahim SA, Wang Z, Spillmann D, Dreier R, Kelsch R, Fischgräbe J, Smollich M, Rossi LH, Sibrowski W, Wülfing P, Kiesel L, Yip GW, Götte M (2009) Differential roles for membrane-bound and soluble syndecan-1 (CD138) in breast cancer progression. Carcinogenesis 30:397–407

Noguer O, Reina M (2009) Is syndecan-2 a key angiogenic element? ScientificWorldJournal 9:729–732

O'Connell MP, Fiori J, Kershner EK, Frank BP, Indig FE, Taub DD, Hoek KS, Weeraratna AT (2009) Heparan sulfate proteoglycan modulation of Wnt5A signal transduction in metastatic melanoma cells. J Biol Chem 284:28704–28712

Oh JH, Kim J, Ahn HJ, Yoon JH, Yoo SC, Choi DS, Lee IS, Ryu HS, Min CK (2009) Syndecan-1 enhances the endometrial cancer invasion by modulating matrix metalloproteinase-9 expression through nuclear factor kappaB. Gynecol Oncol 114:509–515

Orosco A, Fromigue O, Bazille C, Entz-Werle N, Levillain P, Marie PJ, Modrowski D (2007) Syndecan-2 affects the basal and chemotherapy-induced apoptosis in osteosarcoma. Cancer Res 67:3708–3715

Parish CR, Freeman C, Brown KJ, Francis DJ, Cowden WB (1999) Identification of sulfated oligosaccharide-based inhibitors of tumor growth and metastasis using novel in vitro assays for angiogenesis and heparanase activity. Cancer Res 59:3433–3441

Peters MG, Farias E, Colombo L, Filmus J, Puricelli L, de Kier B, Joffé E (2003) Inhibition of invasion and metastasis by glypican-3 in a syngeneic breast cancer model. Breast Cancer Res Treat 80:221–232

Prandoni P, Falanga A, Piccioli A (2005) Cancer and venous thromboembolism. Lancet Oncol 6:401–410

Purushothaman A, Uyama T, Kobayashi F, Yamada S, Sugahara K, Rapraeger AC, Sanderson RD (2010) Heparanase-enhanced shedding of syndecan-1 by myeloma cells promotes endothelial invasion and angiogenesis. Blood 115:2449–2457

Reijmers RM, Groen R, Rozemuller H, Kuil A, de Haan-Kramer A, Csikós T, Martens AC, Spaargaren M, Pals ST (2010) Targeting EXT1 reveals a crucial role for heparan sulfate in the growth of multiple myeloma. Blood 115:601–604

Roy M, Marchetti D (2009) Cell surface heparan sulfate released by heparanase promotes melanoma cell migration and angiogenesis. J Cell Biochem 106:200–209

Savorè C, Zhang C, Muir C, Liu R, Wyrwa J, Shu J, Zhau HE, Chung LW, Carson DD, Farach-Carson MC (2005) Perlecan knockdown in metastatic prostate cancer cells reduces heparin-binding growth factor responses in vitro and tumor growth in vivo. Clin Exp Metastasis 22:377–390

Schlessinger J (2000) Cell signaling by receptor tyrosine kinases. Cell 103:211–225

Sebestyén A, Totth A, Mihalik R, Szakács O, Paku S, Kopper L (2000) Syndecan-1-dependent homotypic cell adhesion in HT58 lymphoma cells. Tumour Biol 21:349–357

Senay C, Lind T, Muguruma K, Tone Y, Kitagawa H, Sugahara K, Lidholt K, Lindahl U, Kusche-Gullberg M (2000) The EXT1/EXT2 tumor suppressors: catalytic activities and role in heparan sulfate biosynthesis. EMBO Rep 1:282–286

Shimada K, Nakamura M, De Velasco MA, Tanaka M, Ouji Y, Konishi N (2009) Syndecan-1, a new target molecule involved in progression of androgen-independent prostate cancer. Cancer Sci 100:1248–1254

Shimada K, Nakamura M, De Velasco MA, Tanaka M, Ouji Y, Miyake M, Fujimoto K, Hirao K, Konishi N (2010) Role of syndecan-1 (CD138) in cell survival of human urothelial carcinoma. Cancer Sci 101:155–160

Siqueira AS, Gama-de-Souza L, Arnaud MV, Pinheiro JJ, Jaeger RG (2010) Laminin-derived peptide AG73 regulates migration, invasion, and protease activity of human oral squamous cell carcinoma cells through syndecan-1 and beta1 integrin. Tumour Biol 31:46–58

Spivak-Kroizman T, Lemmon M, Dikic I, Ladbury JE, Pinchasi D, Huang J, Jaye M, Crumley G, Schlessinger J, Lax I (1994) Heparin-induced oligomerization of FGF molecules is responsible for FGF receptor dimerization, activation, and cell proliferation. Cell 79:1015–1024

Su G, Blaine S, Qiao D, Friedl A (2007) Shedding of syndecan-1 by stromal fibroblasts stimulates human breast cancer cell proliferation via FGF2 activation. J Biol Chem 282:14906–14915

Sutton A, Friand V, Brulé-Donneger S, Chaigneau T, Ziol M, Sainte-Catherine O, Poiré A, Saffar L, Kraemer M, Vassy J, Nahon P, Salzmann JL, Gattegno L, Charnaux N (2007) Stromal cell-derived factor-1/chemokine (C-X-C motif) ligand 12 stimulates human hepatoma cell growth, migration, and invasion. Mol Cancer Res 5:21–33

Suzuki Y, Sakai K, Ueki J, Xu Q, Nakamura T, Shimada H, Nakamura T, Matsumoto K (2010) Inhibition of Met/HGF receptor and angiogenesis by NK4 leads to suppression of tumor growth and migration in malignant pleural mesothelioma. Int J Cancer 127(8):1948–1957

Tímár J, Diczhazhi C, Bartha I, Pogány G, Paku S, Rásó E, Tóvári J, Ladányi A, Lapis K, Kopper L (1995) Modulation of heparan-sulphate/chondroitin-sulphate ratio by glycosaminoglycan biosynthesis inhibitors affects liver metastatic potential of tumor cells. Int J Cancer 62: 755–761

Umezu T, Shibata K, Shimaoka M, Kajiyama H, Yamamoto E, Ino K, Nawa A, Senga T, Kikkawa F (2010) Gene silencing of glypican-3 in clear cell carcinoma of the ovary renders it more sensitive to the apoptotic agent paclitaxel in vitro and in vivo. Cancer Sci 101:143–148

Vuoriluoto K, Jokinen J, Kallio K, Salmivirta M, Heino J, Ivaska J (2008) Syndecan-1 supports integrin alpha2beta1-mediated adhesion to collagen. Exp Cell Res 314:3369–3381

Wang H, Leavitt L, Ramaswamy R, Rapraeger AC (2010) Interaction of syndecan and alpha6-beta4 integrin cytoplasmic domains: regulation of ErbB2-mediated integrin activation. J Biol Chem 285:13569–13579

Xu Y, Liu Y, Yu Q (2004) Angiopoietin-3 is tethered on the cell surface via heparan sulfate proteoglycans. J Biol Chem 279:41179–41188

Yip GW, Smollich M, Götte M (2006) Therapeutic value of glycosaminoglycans in cancer. Mol Cancer Ther 5:2139–2148

Zittermann SI, Capurro M, Shi W, Filmus J (2010) Soluble glypican 3 inhibits the growth of hepatocellular carcinoma in vitro and in vivo. Int J Cancer 126:1291–1301

Zong F, Fthenou E, Wolmer N, Hollósi P, Kovalszky I, Szilák L, Mogler C, Nilsonne G, Tzanakakis G, Dobra K (2009) Syndecan-1 and FGF-2, but not FGF receptor-1, share a common transport route and co-localize with heparanase in the nuclei of mesenchymal tumor cells. PLoS One 4:e7346

Chapter 7
Structure and Biological Activities of Glycosaminoglycan Analogs from Marine Invertebrates: New Therapeutic Agents?

Eliene O. Kozlowski, Angélica M. Gomes, Christiane Sobral Silva, Mariana Sá Pereira, Ana Cristina E.S. de Vilela Silva, and Mauro S.G. Pavão

Abstract Although the oceans are habited by about 230,000 different animal species, only 1% of these organisms have been studied so far. Thus, the sea constitutes an immense reservoir of unique natural molecules with potential biological interest. Among the countless described compounds of marine origin, the sulfated glycosaminoglycans are a particularly interesting group of molecules. These polymers are widely distributed among different marine phyla, occurring in elevated quantities. Normally, the glycosaminoglycans from marine sources contain a high negative charge density due to the presence of sulfate groups in different positions. In addition, unique structural motifs, such as glucose and sulfated fucose branches, rarely described in glycosaminoglycans from terrestrial organisms, are also found. Various pharmacological properties ranging from anticoagulant and antithrombotic to antimetastatic and anti-inflammatory have been attributed to these molecules. The interaction and modulation of plasma coagulation proteases and inhibitors, adhesion molecules, and growth factors, among others, are the basis for their pharmacological effects. In this chapter, we review the phylogenetic distribution, the structure, and the biological effects of the marine glycosaminoglycans, as well

E.O. Kozlowski, A.M. Gomes, and C.S. Silva
Programa de Glicobiologia, Instituto de Bioquímica Médica and Hospital Universitário Clementino Fraga Filho, Universidade Federal do Rio de Janeiro, Cidade Universitaria, Rio de Janeiro, Brazil

M.S. Pereira and A.C.E.S. de Vilela Silva
Instituto de Ciências Biomédicas, Universidade Federal do Rio de Janeiro, Centro de Ciências da Saúde, Cidade Universitaria, Rio de Janeiro, Brazil

M.S.G. Pavão (✉)
Programa de Glicobiologia, Instituto de Bioquímica Médica and Hospital Universitário Clementino Fraga Filho, Universidade Federal do Rio de Janeiro, Cidade Universitaria, Rio de Janeiro, Brazil
and
Laboratório de Tecido Conjuntivo, Hospital Universitário Clementino Fraga Filho, Cidade Universitária, Rua Rodolpho Paulo Rocco 255, 4° andar, sala 4A-08, CEP 21941-913 Rio de Janeiro, Brazil
e-mail: mpavao@hucff.ufrj.br

M.S.G. Pavão (ed.), *Glycans in Diseases and Therapeutics*, Biology of Extracellular Matrix, 159
DOI 10.1007/978-3-642-16833-8_7,

as the molecular mechanisms involved in some of their biological activities. We also briefly discuss the possibility of using these glycans as therapeutic agents.

Abbreviations

GalNAc	*N*-acetyl galactosamine
GalNAc4S	4-O-sulfated *N*-acetyl galactosamine
GalNAc4S, 6S	4-O- and 6-O-sulfated *N*-acetyl galactosamine
GalNAc6S	6-O-sulfated *N*-acetyl galactosamine
GlcA	Glucuronic acid
GlcA2S	2-O-sulfated glucuronic acid
GlcA3S	3-O-sulfated glucuronic acid
GlcNAc	*N*-acetyl glucosamine
GlcNS	N-sulfated glucosamine
GlcNS, 3S, 6S	N-, 3-O-, and 6-O-sulfated glucosamine
GlcNS, 6S	N- and 6-sulfated glucosamine
HexA	Hexuronic acid
IdoA	Iduronic acid
IdoA2S	2-O-sulfated iduronic acid

7.1 Diversity of Sea-Life and Unique Biological Molecules

The oceans cover about 70% of the earth's surface and shelter an incredible diverse web of life, with species from almost all groups of living animals, including some exclusive marine phyla. The Census of Marine Life (http://www.coml.org/census) coordinated by The Consortium for Ocean Leadership in Washington DC, USA, estimates that about 230,000 species of marine animals have been described.

During the course of evolution, the marine organisms have adapted to extreme environmental conditions, such as high salt concentration, extreme temperatures, high pressure, and low nutrient concentrations. As a result, biological molecules with unique structural features not present in their counterparts found in terrestrial organisms have arisen, providing an enormous potential for different biotechnological applications in industry and medicine (Haefner 2003). While there are a few marine natural products currently on the market or in clinical trials, marine organisms represent the greatest unexploited source of potential pharmaceuticals.

In addition to secondary metabolites that constitute the driving force of the marine natural products area (Molinski 2009), marine invertebrate animals also synthesize macromolecules, such as sulfated polysaccharides, of potential biotechnological application. These polymers can be divided into two categories: the sulfated fucans from sea cucumber and sea urchins and sulfated galactans from ascidians and sea urchins (Pomin 2009; Pomin and Mourao 2008), and a group of compounds with structures similar to the vertebrate glycosaminoglycans, but containing distinct sulfation patterns and/or the occurrence of neutral sugars as branched units (Tables 7.1 and 7.2).

7.2 Glycosaminoglycans from Marine Invertebrate Phyla and Their Biological Properties

Glycosaminoglycan-like polysaccharides are present in several families of invertebrate animals from the phyla Cnidaria, Arthropoda, Mollusca, Echinodermata, and Chordata (Tables 7.1 and 7.2). In general, the invertebrate glycosaminoglycans possess higher charge densities, especially the galactosaminoglycans, and sulfation patterns not commonly observed in their vertebrate counterparts. An exclusive characteristic of some of these glycans is the occurrence of branches of neutral sugars that can contain a sulfate group.

Cnidaria is a poorly documented phylum regarding the presence of glycosaminoglycan-like polysaccharides. The occurrence of a glucosaminoglycan was described in only one species of the class Anthozoa. The polysaccharide isolated from the species *Nematostella vectensis* (starlet sea anemone) has been characterized as a heparan sulfate with unusual disaccharide composition (Feta et al. 2009). To date, no galactosaminoglycan has been described in this phylum.

In the phylum Arthropoda, an oversulfated chondroitin sulfate has been described in only one merostomata species of the family Liminulidae (Sugahara et al. 1996; Fongmoon et al. 2007) (Table 7.1). Heparin/heparan sulfate polymers, on the other hand, are more highly distributed, having been described in four species from three families of decapods (Table 7.2) (Cassaro and Dietrich 1977; Hovingh and Linker 1982; Dietrich et al. 1999; Demir et al. 2001; Brito et al. 2008) and in one brachiopod species of the family Artemiidae (Chavante et al. 2000; Nader et al. 2001).

In mollusks, oversulfated chondroitin sulfates are present in three families of cephalopods of the order Teuthida (Mathews et al. 1962; Kawai et al. 1966; Habuchi et al. 1971, 1977; Karamanos et al. 1991; Kinoshita et al. 1997; Clement et al. 1999; Deepa et al. 2002; Ueoka et al. 2000; Kawashima et al. 2002; Kinoshita-Toyoda et al. 2004; Bergefall et al. 2005) and in one bivalve species of the family Arcidae (Volpi and Maccari 2009) (Table 7.1). Heparinoids, mostly of the heparin type, are restricted to the class of Bivalvia (Table 7.2), occurring in several species of the family Veneridae (Dietrich et al. 1985, 1989; Jordan and Marcum 1986; Pejler et al. 1987; Ulrich and Boon 2001; Santos et al. 2002; Cesaretti et al. 2004; Luppi et al. 2005; Vijayabaskar et al. 2008), and in different families of the order Veneroida (Dietrich et al. 1989; Burson et al. 1956; Nader et al. 1984; Arumugam et al. 2008; Vijayabaskar and Somasundaram 2009). Heparin-like glycans also

Table 7.1 Characteristics of galactosaminoglycans found in marine invertebrates

Phylum	Class	Order	Family	Species	Tissue	Gag (Disaccharide units)	Biological activity	References
Arthropoda	Merostomata	Xiphosura	Limulidae	*Tachypleus tridentatus*	Gill cartilage	Oversulfated CS [GlcA3S–GalNAc4S] [GlcA3S–GalNAc6S] [GlcA3S–GalNAc4S,6S] (fucose branches)	Neurutogenic activity	Fongmoon et al. (2007), Nobuko and Murakami (1982)
Mollusca	Chephalopoda	Teuthida	Ommastrephidae	*Illex illecebrosus*	Cornea	Oversulfated CS [GlcA–GalNAc4S] (52%) [GlcA2S–GalNAc6S] (28%) (glucose branches)	ND	Karamanos et al. (1991)
					Skin	CS [GlcA–GalNAc4S] (49%) [GlcA–GalNAc6S] (20%) [GlcA–GalNAc] (30%) (glucose branches)	ND	Karamanos et al. (1988, 1994), Aletras et al. (1991), Karamanos (1992)
			Loliginidae	*Loligo sp*	Cranial cartilage	Oversulfated CS [GlcA–GalNAc4S,6S] (major) [GlcA3S–GalNAc4S] (10%) [GlcA3S–GalNAc6S] (10%) [GlcA3S–GalNAc4S,6S] (10%) (glucose branches)	Neuronal migration inhibitory activity; neuronal cell adhesion inhbitiory activity; P- and L-selectin inhibitory activity; neurite outgrowth; inhibition of herpes simplex virus infectivity	Karamanos et al. (1991), Habuchi et al. (1971, 1977), Kinoshita et al. (1997), Clement et al. (1999), Deepa et al. (2002), Ueoka et al. (2000), Kawashima et al. (2002), Kinoshita-Toyoda et al. (2004), Bergefall et al. (2005), Mathews et al. (1962)
			Thysanoteuthidae	*Thysanoteuthis rhombus*	Fin	Oversulfated CS [GlcA–GalNAc] (77%) [GlcA–GalNAc4S] (8%) [GlcA–GalNAc4S,6S] (11%) [GlcA–GalNAc6S] (4%)	ND	Tamura et al. (2009)

				Arms	Oversulfated CS [GlcA–GalNAc] (88%) [GlcA–GalNAc4S] (6%) [GlcA–GalNAc6S] (2%) [GlcA–GalNAc4S,6S] (4%)	ND	Tamura et al. (2009)
				Skin	Oversulfated CS [GlcA–GalNAc] (17%) [GlcA–GalNAc4S] (26%) [GlcA–GalNAc6S] (19%) [GlcA–GalNAc4S6S] (38%)	ND	Tamura et al. (2009)
				Eyes	Oversulfated CS [GlcA–GalNAc] (13%) [GlcA–GalNAc4S] (52%) [GlcA–GalNAc6S] (13%) [GlcA–GalNAc4S,6S] (22%)	ND	Tamura et al. (2009)
				Mantle	Oversulfated CS [GlcA–GalNAc] (4%) [GlcA–GalNAc4S] (54%) [GlcA–GalNAc6S] (16%) [GlcA–GalNAc4S,6S] (26%)	ND	Tamura et al. (2009)
		NS	NS	Liver integument	CS [GlcA–GalNAc4S] (41.8%) [GlcA–GalNAc4S,6S] (21.6%) [GlcA3S–GalNAc4S](5%) [GlcA3S–GalNAc6S] (ND) [GlcA3S–3GalNAc4S,6S] (2.8%)	Neurite outgrowth-promoting activity	Shetty et al. (2009)
Bivalvia	Arcoida	Arcidae	*Scapharca inaequivalvis*	Internal organs	Oversulfated DS [IdoA–GalNAc4S] (major) [IdoA2S–GalNAc4S] [HexA–GalNAc] (minor)	Heparin cofactor II activity	Volpi and Maccari (2009)

(*continued*)

Table 7.1 (continued)

Phylum	Class	Order	Family	Species	Tissue	Gag (Disaccharide units)	Biological activity	References
Echinodermata	Echinoidea	Echinoida	Strongylocentrotidae	*Hemicentrotus pulcherrimus*	Embryo	DS	ND	Oguri and Yamagata (1978)
				Strongylocentrotus purpuratus	Embryo (gastrula)	Oversulfated DS [IdoA–GalNAc4S,6S]	ND	Vilela-Silva et al. (2001)
		Clypeasteroida	Clypeasteridae	*Clypeaster japonicus*	Embryo (gastrula)	DS [IdoA–GalNAc6S]	ND	Yamaguchi et al. (1989)
	Holothuroidea	Aspidochirotida	Holothuridae	*Holothuria grisea* (*Ludwigothurea grisea*)	Body wall	Oversulfated CS [GlcA–GalNAc4S,6S] (GlcA3S residues) (sulfated fucose branches)	Anticoagulant and antithrombotic activity; anti P- and L-selectin inhibitory activity; antimetastatic activity; inhibition of smooth muscle cell proliferation; induces FGF-mediated endothelial cell proliferation; induces TFPI release; induces plasminogen activation by t-PAs and U-PA; induces FGF-2-mediated endothelial cell angiogenesis	Vieira and Mourao (1988), Vieira et al. (1991), Mourao et al. (1996, 2001), Zancan and Mourao (2004), Borsig et al. (2007), Fonseca and Mourao (2006), Glauser et al. (2008), Buyue and Sheehan (2009), Tapon-Bretaudiere et al. (2000)
			Stichopodidae	*Apostichopus japonicus* (*Stichopus japonicus*)	Body wall	Oversulfated CS [GlcA–GalNAc4S,6S] (GlcA3S residues) (sulfated fucose branches)	Induces plasminogen activation by t-Pas	Kariya et al. (1990, 1997, 2002)
Chordata	Ascidiacea	Phlebobranchia	Ascidiidae	*Phallusia nigra* (*Ascida nigra*)	Internal organs	Oversulfated DS [IdoA2S–GalNAc6S] (80%) [IdoA–GalNAc6S]	Neurotrophic factor-binding activity; neurite outgrowth-promoting activity	Pavao et al. (1995, 1998), Mourao et al. (1997), Bao et al. (2005)

	Stolidobranchia	Pyuridae	*Halocynthia pyriformis*	Internal organs	Oversulfated DS [IdoA2S–GalNAc4S]	Anticoagulant activity; heparin cofactor II activity	Bao et al. (2005)
		Styelidae	*Styela clava*	Tunic	CS	Anti-inflammatory	Xu et al. (2008a, b)
			Styela plicata	Intestine, heart, mantle, pharynx	Oversulfated DS [IdoA2S–GalNAc4S] (70%) [IdoA–GalNAc4S]	Anticoagulant and antithrombotic activities; heparin cofator II activity; modest neurite outgrowth-promoting activity	Pavao et al. (1994, 1998), Gandra et al. (2000, 2006), Hikino et al. (2003), Vicente et al. (2001)
Myxini	Myxiniformes	Myxinidae	*Eptatretus burgeri*	Notochord	Oversulfated hibrid CS/DS [GlcA/IdoUA–GalNAc 4,6S]	Neurite outgrowth-promoting activity	Nandini et al. (2004)

Table 7.2 Characteristics of heparinoids found in marine invertebrates

Phylum	Class	Order	Family	Species	Tissue	Gag (Disaccharide units)	Biological activity	References
Cnidaria	Anthozoa	Actiniaria	Edwardsiidae	*Nematostella vectensis*	Embryo (mixed stages)	Heparan sulfate [IdoA2S–GlcNS3S] (40%) [IdoA2S–GlcNS6S] [IdoA2S–GlcNS] [IdoA–GlcNS6S]	ND	Feta et al. (2009)
Arthropoda	Malacostraca	Decapoda	Ocypodidae	*Ucides cordatus*	Internal organs	Heparin [HexA–GlcNS,6S] (major) [HexA–GlcNS] [HexA2S–GlcNS] [HexA2S–GlcNS,6S]	Anticoagulant activity	Medeiros et al. (2000)
			Nephropidae	*Homarus americanus*	Lobster paste	Oversulfated HS [GlcA–GlcNS,6S] (major)	Low anticoagulant activity	Hovingh and Linker (1982)
			Penaeidae	*Penaeus brasiliensis*	Head	Low-molecular-weight heparin [GlcA–GlcNS,6S] (46%) [GlcA2S–GlcNS,6S] (13%) [GlcA–GlcNS] (11%) [GlcA2S–GlcNS] (16%) [GlcA2S–GlcNAc,6S] (5%)	Anti-Xa and anti-IIa activities; heparin cofactor II activity; antithrombotic activity	Dietrich et al. (1999), Demir et al. (2001)
				Litopenaeus vannamei	Cephalotorax	Heparin [GlcA–GlcNS,6S] and [GlcA2S–GlcNS] (major sequences)	Anticoagulant activity; anti-inflammatory activity; proMMP-2 and MMP-9 inhibitory activity	Brito et al. (2008)
	Branchiopoda	Anostraca	Artemiidae	*Artemia franciscana*	Whole animal	Oversulfated HS [GlcA–GlcNS] and [GlcA–GlcNS,6S] (major sequences)	Low anticoagulant activity; heparin cofactor II activity	Chavante et al. (2000), Nader et al. (2001)
Mollusca	Bivalvia	Veneroida	Veneridae	*Katelysia opima*	ND	Heparin-like	High anticoagulant activity	Vijayabaskar et al. (2008)
				Callista chione	Internal organs	Heparin [IdoA2S–GlcNS,6S] (46%) [GlcA–GlcNS,6S] (11%) [IdoA2S–GlcNS] (15%) [IdoA–GlcNAc6S] (2%) [GlcA–GlcNS3S,6S] (1.4%) [IdoA–GlcNS,6S] (12%)	Anticoagulant activity	Luppi et al. (2005)
				Tapes philippinarum	Internal organs	Heparin [IdoA2S–GlcNS,6S] (80.6%)	High anticoagulant activity	Cesaretti et al. (2004)

				[GlcA–GlcNS,6S] (4%) [IdoA2S–GlcNS] (6%) [IdoA–GlcNAc6S] (2%) [GlcA–GlcNS,3S,6S] (2.3%) [IdoA–GlcNS,6S] (0.2%)		
		Anomalocardia brasiliana	Labial palp, intestine, ctenidium, mantle, and foot tissues	Heparin [GlcA–Glc6S] [IdoA2S–GlcN] [IdoA2S–GlcNS] (major sequences)	ND	Dietrich et al. (1985), Pejler et al. (1987), Santos et al. (2002)
		Mercenaria mercenaria	Labial palp, ctenidium, intestine, siphon, and pallium	Heparin (3-O-sulfated GlcNAc residues)	High anticoagulant activity; anti-IIa and anti-Xa activities	Jordan and Marcum (1986), Ulrich and Boon (2001)
		Tivela mactroides	Internal organs	Heparin 3-O-sulfated sequences [GlcA–GlcNS,6S] (Major) [GlcA2S–GlcNS,6S] [IdoA2S–GlcNS,6S]	High anticoagulant activity	Pejler et al. (1987), Dietrich et al. (1989)
	Arcticidae	*Cyprina islandica (Arctica islandica)*	Internal organs	Heparin-like (mactin)	High anticoagulant activity	Burson et al. (1956)
	Solecurtidae	*Tagelus gibbus (Tagelus plebeius)*	Internal organs	Heparan sulfate [GlcA–GlcNS,6S] (Major) [GlcA–GlcNS]	Very low anticoagulant activity	Nader et al. (1984)
	Mactridae	*Spisula solidissima*	Internal organs	Heparin-like (mactin)	High anticoagulant activity	Burson et al. (1956)
	Tridacnidae	*Tridacna maxima*	Internal organs	Heparin-like	Low anticoagulant activity	Nader et al. (1984)
	Donacidae	*Donnax striatus*	Internal organs	Heparin [GlcA–GlcNS,6S] (Major) [GlcA2S–GlcNS,6S] [IdoA2S–GlcNS,6S]	High anticoagulant activity	Dietrich et al. (1989)
		Donax cuneatus	Internal organs	Heparin-like	Anticoagulant activity	Vijayabaskar and Somasundaram (2009)
Pectonoida	Pectinidae	*Amusium pleuronectes*	Internal organs	Low-molecular-weight heparin (porcine-type)	Anticoagulant activity	Saravanan and Shanmugam (2010a, b)
		Nodipecten nodosus	Internal organs	Heparin-like [GlcA–GlcNAc] (60%) IdoA2*s*-, GlcA2*s*-, Glc-3S-, GlcNS-containing residues	Low anticoagulant activity; high antithrombotic effect	Gomes et al. (2010)

(continued)

Table 7.2 (continued)

Phylum	Class	Order	Family	Species	Tissue	Gag (Disaccharide units)	Biological activity	References
		Mytiloida	Mytilidae	*Perna viridis*	Internal organs	Heparin-like	Anticoagulant activity	Arumugam et al. (2008)
Echinodermata	Echinoidea	Clypeasteroida	Mellitidae	*Mellita quinquisperforata*	Internal organs	Heparin-like [IdoA2S–GlcNS,6S] (Major)	Anticoagulant activity	Demir et al. (2001), Medeiros et al. (2000)
Chordata	Ascidiacea	Stolidobranchia	Styelidae	*Styela plicata*	Intestine	Heparin [HexA–GlcNS,6S] (39%) [IdoA2S–GlcNS,6S] (61%)	Anticoagulant activity	Gandra et al. (2000)
					Heart	Heparin [HexA–GlcNS,6S] (31%) [IdoA2S–GlcNS,6S] (69%)	Anticoagulant activity	Gandra et al. (2000)
					Pharynx	Heparin [HexA–GlcNS6S] (25.5%) [IdoA2S–GlcNS,6S] (74.5%)	Anticoagulant activity	Gandra et al. (2000)
					Mantle	Heparin [HexA–GlcNS,6S] (58.5%) [IdoA2S–GlcNS,6S] (41.5%)	Anticoagulant activity	Gandra et al. (2000)
					Test cells (vertebrate mast cell-like)	Heparin [IdoA2S–GlcNS,6S] (75%) [HexA–GlcNS6S] (25%)	Anticoagulant and antithrombotic activities	
					Hemolymph basophil-like cells	Heparin [IdoA2S–GlcNS]; IdoA2S–GlcNS,6S]; [IdoA2S– GlcNS,3S]; [IdoA2S–GlcNS,3S,6S]	ND	Cavalcante et al. (2000, 2002); de Barros et al. (2007)
					Whole body	Heparin [IdoA2S–GlcNS,6S] (38.3%) [HexA–GlcNS6S] (2%) [IdoA2S–GlcNS] (47.5%) [HexA–GlcNAc6S] (1.4%) [IdoA2S–GlcNS3S] (8%) [IdoA2S–GlcNS3S6S] (2.8%)	Anticoagulant and antithrombotic activities; anti-inflammatory activity; (inhibition of TGF-β, VEGF, and TNF-α production in inflammed intestine)	Santos et al. (2007), Belmiro et al. (2009)

ND not determined; *NS* not specified by the authors; *DS* dermatan sulfate; *CS* chondroitin sulfate occur in families of the orders Pectonoida (Saravanan and Shanmugam 2010b) and Mytiloida (Arumugam et al. 2008) (Table 7.1).

Oversulfated galactosaminoglycans are present in two classes of echinoderms: Echinoidea and Holothuroidea (Table 7.1). Dermatan sulfate (Oguri and Yamagata 1978; Yamaguchi et al. 1989) and oversulfated dermatan sulfate (Vilela-Silva et al. 2001) have been described in three echinoid species of the families Strongylocentrotidae and Clypeasteridae. Holothurians do not synthesize dermatan sulfate but unique oversulfated chondroitin sulfates containing sulfated fucose branches. These glycans have been described in the families Holothuridae (Vieira and Mourao 1988; Vieira et al. 1991) and Stichopodidae (Kariya et al. 2002). The occurrence of heparin-like glycans is restricted to only one family of echinoid: the Mellitidae (Medeiros et al. 2000; Demir et al. 2001) (Table 7.2).

In invertebrate chordates, galactosaminoglycans with high degree of sulfation have been described in the classes: Ascidiacea and Myxini (Table 7.1). Oversulfated dermatan sulfates occur in one species of phlebobranchs (Pavao et al. 1995, 1998; Mourao et al. 1997) and in three species of stolidobranch ascidians (Pavao et al. 1998, 1994). Chondroitin sulfate and a hybrid chondroitin/dermatan sulfate polymer have been described in one ascidian species of the family Styelidae (Xu et al. 2008b) and in one myxini species of the family Myxinidae (Nandini et al. 2004), respectively.

A heparin-like glycan has been described only in one species of stolidobranchia ascidian from the family Styelidae (Cavalcante et al. 2000; Gandra et al. 2000, 2006). No heparin-like polymer has been described in the class Myxini (Table 7.2).

7.2.1 Cnidaria

7.2.1.1 Galactosaminoglycans

There have been no studies on the characterization of galactosaminoglycans from marine Cnidarians. However, a chondroitin sulfate with low degree of sulfation has previously been shown to be present in *Hydra*, a freshwater invertebrate of the class Hydrozoa (Yamada et al. 2007).

7.2.1.2 Glucosaminoglycans

An oversulfated heparan sulfate with a unique structure was isolated from the embryos of the anthozoan *N. vectensis*. This heparinoid contains a high amount of the unusual disaccharide [IdoA2S–GlcNS,3S], corresponding to ~40% of the total disaccharide units (Feta et al. 2009).

7.2.2 *Arthopoda*

7.2.2.1 Galactosaminoglycans

A chondroitin sulfate with unique GlcA3S-containing disulfated disaccharide units, [GlcA3S–GalNAc4S], [GlcA3S–GalNAc6S], and [GlcA3S–GalNAc4S,6S], in addition to the common monosulfated disaccharide unit [GlcA–GalNAc4S] was isolated from the gill cartilage of the merostomata *Tachypleus tridentatus* (king crab or horseshoe crab) (Table 7.1). The crab cartilage chondroitin sulfate was denominated by chondroitin sulfate K. The units [GlcA3S–GalNAc4S], [GlcA3S–GalNAc6S], and [GlcA3S–GalNAc4S,6S] were denominated as K-unit, L-unit, and M-unit, respectively (Sugahara et al. 1996; Nobuko and Murakami 1982). The presence of fucose branches has also been reported to occur in chondroitin sulfate K (Sugahara et al. 1996).

Interestingly, GlcA3S has been detected in human peripheral nerves using the monoclonal antibody HNK-1, raised against human natural killer cells (Chou et al. 1986; Ariga et al. 1987), suggesting that chondroitin sulfate K may occur on proteoglycans in the nervous system. In fact, this glycan has strong affinity toward growth/neurotrophic factors expressed in the brain during development, possessing stronger binding affinity to FGF-18, HGF, PTN, and MK, as well as a potent neurite outgrowth-promoting activity on mouse hippocampal neurons (Fongmoon et al. 2007).

7.2.2.2 Glucosaminoglycans

Heparin-like glycosaminoglycans occur in two classes of arthropods: Malacostraca and Brachiopoda. Heparins with different disaccharide composition and anticoagulant activities have been described in three families of decapods (Table 7.2). The heparinoids isolated from the decapods *Ucides cordatus* (crab), *Homarus americanus* (lobster), *Penaeus brasiliensis* (pink shrimp), and *Litopeneaus vannamei* (white leg shrimp) are composed of the same major disaccharide unit [HexA–GlcNS,6S] (Hovingh and Linker 1982; Dietrich et al. 1999; Demir et al. 2001; Brito et al. 2008; Medeiros et al. 2000). The lobster heparinoid is an oversulfated heparan sulfate and has a low anticoagulant activity (Hovingh and Linker 1982), as also observed for the crab heparinoid (Medeiros et al. 2000). The heparin from the white leg shrimp shows reduced anticoagulant and hemorrhagic activities and, nevertheless, has pronounced anti-inflammatory, antimetalloproteinase, and antiangiogenic activities (Brito et al. 2008; Dreyfuss et al. 2010), emerging as a possible candidate drug for treating angioproliferative and inflammatory conditions. The heparin from pink shrimp has a low molecular weight, possessing antithrombin-mediated anti-IIa and anti-Xa activities. It also activates heparin cofactor II and has antithrombotic activity (Dietrich et al. 1999; Demir et al. 2001).

The brachiopod *Artemia franciscana* (brine shrimp) contains an oversulfated heparan sulfate consisting of [GlcA–GlcNS] and [GlcA–GlcNS,6S] disaccharide units. It has a low anticoagulant activity and potentiates heparin cofactor II (Chavante et al. 2000; Nader et al. 2001).

7.2.3 Mollusca

7.2.3.1 Galactosaminoglycans

Oversulfated chondroitin sulfates have been described in the three families of cephalopod mollusks: Ommastrephidae, Loliginidae, and Thysanoteuthidae. The chondroitin sulfate from the cornea of the ommastrephid *Illex illecebrosus* (northern short fin squid) contains a glucose-branched oversulfated chondroitin sulfate, consisting of [GlcA–GalNAC4S] (major) and [GlcA2S–GalNAC6S] (D-unit) disaccharide units (Karamanos et al. 1991). On the other hand, the skin of this cephalopod has a glucose-branched chondroitin sulfate composed by [GlcA–GalNAC4S] (A-unit) (major), [GlcA–GalNAC6S] (C-unit), and [GlcA–GalNAC] (Karamanos et al. 1988, 1994; Aletras et al. 1991; Karamanos 1992).

A highly studied glucose-branched oversulfated chondroitin sulfate has been isolated from the cranial cartilage of the loliginid *Loligo sp.* (squid). The major disaccharide unit of this oversulfated chondroitin is [GlcA–GalNAc4S,6S] (E-unit). The polymer also contains minor amounts of A- and C-unit, as well as the trisulfated disaccharide unit [GlcA3S–GalNAc4S,6S] (M-unit) (10%) (Kawai et al. 1966; Habuchi et al. 1971, 1977; Kinoshita et al. 1997; Mathews et al. 1962). About 5 mol% of all disaccharide units are substituted by glucose (Kinoshita-Toyoda et al. 2004). Another chondroitin sulfate enriched in E-units and GlcA3S-containing disaccharide units (K-, L-, and M-unit) has been isolated from squid liver integument (Shetty et al. 2009).

Several biological activities have been described for this chondroitin sulfate E: neuronal migration inhibitory activity, neuronal cell adhesion inhibitory activity, P- and L-selectin inhibitory activity, neurite outgrowth, and inhibition of herpes simplex virus infectivity (Clement et al. 1999; Deepa et al. 2002; Ueoka et al. 2000; Kawashima et al. 2002; Bergefall et al. 2005; Shetty et al. 2009). The effect of the squid chondroitin sulfate E on the nervous system seems to be related to the expression and/or modulation of neuroregulatory function of nervous growth factors, including MK, which has a strong affinity binding to the glycan (Ueoka et al. 2000). The inhibition of proinflammatory molecules, such as P- and L-selectins, and various chemokines occurs through E-unit-containing sequences in the squid chondroitin sulfate E (Kawashima et al. 2002). The inhibition of infection of herpes simplex virus 1 by the mollusk chondroitin sulfate E occurs through the binding of glycoprotein C to the glycan chains on the host mammalian cells (Bergefall et al. 2005).

Chondroitin sulfates with the same common core structure [GlcA–GalNAc] with various sulfation profiles have been described in different organs (fin, arms, skin, eyes, and mantle) of the thysanoteuthid mollusk *Thysanoteuthis rhombus* (diamond squid) (Tamura et al. 2009). The chondroitin sulfate in the organs differs on the percentage of the following disaccharides units [GlcA–GalNAc], [GlcA– GalNAc4S], [GlcA–GalNAc6S], and [GlcA–GalNAc4S,6S]. In the skin, eyes, and mantle, the preponderant unit is the disulfated disaccharide [GlcA–GalNAc4S,6S]. The nonsulfated disaccharide [GlcA–GalNAc] prevails in the fin and arms (Tamura et al. 2009).

Interestingly, no chondroitin sulfate has been detected in the class Bivalvia. An oversulfated dermatan sulfate enriched in [IdoA–GalNAc4S] (major) and [IdoA2S–GalNAc4S] disaccharide units, possessing heparin cofactor II activity, has been described in the internal organs of the arcoid *Scapharca inaequivalvis* (Ark clam) (Volpi and Maccari 2009).

7.2.3.2 Glucosaminoglycans

Heparin-like glycosaminoglycans occur in three orders of bivalve mollusks: Veneroida, Pectonoida, and Mytiloida. The family Verenidae contains the major number of species where heparin-like glycans have been detected (Table 7.2). In general, the heparins from these species are very heterogeneous in terms of disaccharide composition, containing different percentages of the following units: [IdoA2S–GlcNS,6S], [GlcA–GlcNS,6S], [IdoA2S–GlcNS], [IdoA–GlcNAc6S], [GlcA–GlcNS,3S,6S], and [IdoA–GlcNS6S]. The 3-O-sulfated GlcNAc residue, required for high-affinity binding to antithrombin (Lindahl et al. 1980), occurs in higher amounts in the heparins with higher anticoagulant activity (Dietrich et al. 1985, 1989; Jordan and Marcum 1986; Pejler et al. 1987; Ulrich and Boon 2001; Santos et al. 2002; Cesaretti et al. 2004; Luppi et al. 2005; Vijayabaskar et al. 2008), for example, those from *Katelysia opima* (clam) (Vijayabaskar et al. 2008), *Tapes philippinarum* (manila clam) (Cesaretti et al. 2004), *Mercenaria mercenaria* (hard clam) (Jordan and Marcum 1986; Pejler et al. 1987), and *Tivela mactroides* (trigonal tivela) (Pejler et al. 1987; Dietrich et al. 1989) (Table 7.2).

Two species of the family Pectinidae contain heparin-like polymers with interesting structural characteristics (Table 7.2). The heparinoid from the pectonoid *Amusium pleuronectes* (Asian moon scallop) has low molecular weight and a chemical composition similar to that of porcine intestinal heparin (Saravanan and Shanmugam 2010b). This heparinoid possesses anticoagulant activity and a protective effect against myocardial injury in a model of myocardial infarction (Saravanan and Shanmugam 2010a). On the other hand, the heparinoid from the pectonoid *Nodipecten nodosus* (lion's paw scallop) is a hybrid heparin/heparan sulfate polymer, containing GlcA2s and Glc-3S residues. It has a low anticoagulant activity and a potent antithrombotic activity on an arterial model of thrombosis (Gomes et al. 2010).

Heparinoids with different disaccharide composition, mostly similar to vertebrate heparins with various anticoagulant activities, are also present in different families of the order Veneroida (Table 7.2) (Dietrich et al. 1989; Burson et al. 1956; Nader et al. 1984; Arumugam et al. 2008; Vijayabaskar and Somasundaram 2009).

7.2.4 *Echinodermata*

7.2.4.1 Galactosaminoglycans

Two orders of echinoderms, Echinoidea and Holothuroidea, contain dermatan sulfate and chondroitin sulfate, respectively, with highly sulfated disaccharide units (Table 7.1). The embryos of the echinoids *Hemicentrotus pulcherrimus* (sea urchin) and *Strongylocentrotus purpuratus* (purple sea urchin) contain dermatan sulfates with different sulfation patterns. While *S. purpuratus* dermatan sulfate is composed by the disulfated disaccharide [IdoA–GalNAc4S,6S] (Vilela-Silva et al. 2001), that of *H. pulcherrimus* is formed by the common monosulfated disaccharide [IdoA–GalNAc4S] (Oguri and Yamagata 1978). Interestingly, the embryos of the clypeasteroid *Clypeaster japonicus* are composed by the unusual disaccharide [IdoA–GalNAc6S] (Yamaguchi et al. 1989). The DS from *S. purpuratus* embryos, probably due to its high content of iE-units, exerts a high neurite outgrowth-promoting activity in mouse hippocampal neurons (Hikino et al. 2003).

Unique oversulfated chondroitin sulfates containing sulfated fucose branches and unusual GlcA3S residues have been described in the body wall of the holothurians: *Holothuria grisea* (gray sea cucumber) (Vieira and Mourao 1988; Vieira et al. 1991) and *Apostichopus japonicus* (Japanese sea cucumber) (Kariya et al. 1990, 1997, 2002) (Table 7.1). The *H. grisea* glycan has a core like that of mammalian chondroitin sulfate E [GlcA–GalNAc4S,6S], but substituted at the 3-position of the GlcAc residues with sulfated fucose branches (Vieira and Mourao 1988; Vieira et al. 1991). The sulfated fucose residues can occur as fucose-4S, fucose-2,4, and fucose-3S,4S. The fucose residues are concentrated toward the nonreducing end of the polysaccharide chains (Mourao et al. 1996). The *A. japonicus* chondroitin also consists of a central core of chondroitin sulfate E substituted at the 3-position of the GlcAc residues or at the 4- or 6-position of the GalNAc with sulfated fucose branches in almost all disaccharide units (Kariya et al. 1990, 1997).

Several biological properties have been reported for the fucosylated chondroitin sulfate from sea cucumbers (Table 7.1). The most studied of those activities are the anticoagulant and antithrombotic that revealed a new molecular mechanism involved in the antithrombotic effect of sulfated glycosaminoglycans (Mourao et al. 1996, 2001; Zancan and Mourao 2004; Borsig et al. 2007; Fonseca and Mourao 2006; Glauser et al. 2008; Buyue and Sheehan 2009).

The fucosylated chondroitin sulfates have high anticoagulant and antithrombotic activities (Mourao et al. 1996, 2001; Zancan and Mourao 2004; Glauser et al. 2008; Buyue and Sheehan 2009) (Table 7.1). Two anticoagulant mechanisms have been described for these glycans: a serpin-dependent, which is mostly due to activating thrombin inhibition by heparin cofactor II, and a serpin-independent mechanism, which involves inhibition of factor-Xa and thrombin generation by the tenase and prothrombinase complexes, respectively (Mourao et al. 2001; Glauser et al. 2008; Buyue and Sheehan 2009). The inhibition of the tenase complex involves binding of the fucosylated chondroitin sulfate to the factor-IXa heparin-binding exosite

(Buyue and Sheehan 2009). Studies on the structure versus biological activity indicated that removal of the sulfated fucose branches, but not reduction of the glucuronic acid residues to glucose, abolished the anticoagulant and antithrombotic activities of the sea cucumber glycan (Mourao et al. 1996; Glauser et al. 2008).

The fucosylated chondroitin sulfate from *H. grisea* has distinct effects on thrombus formation in vein and artery after intravascular administration. Thus, inhibition of thrombosis in artery occurs at low doses and does not modify the anticoagulant activity of the plasma. In vein, on the contrary, the antithrombotic activity of the fucosylated chondroitin requires high doses and occurs concomitantly with an increase in the anticoagulant activity of the plasma (Zancan and Mourao 2004). Interestingly, the *H. grisea* glycan induced a dose-dependent increase in the plasma anticoagulant activity after oral administration to rats. Additionally, animals receiving daily oral doses of this glycosaminoglycan showed a decrease in thrombus weight on experimental models of venous and arterial thrombosis (Fonseca and Mourao 2006).

In addition to its ability to inhibit thrombus formation in artery and veins, the fucosylated chondroitin sulfate from sea cucumber also inhibits smooth muscle cell proliferation and, in the presence of fibroblast growth factor-2, increases the capacity of endothelial cells to form vascular tubes in a well-organized capillary-like network (Tapon-Bretaudiere et al. 2000). Overall, these activities make this new glycosaminoglycan a potential therapeutic drug in pathological conditions affecting blood vessels such as the neovascularization of ischemic areas.

In a recent work, Borsig et al. (2007) showed that the fucosylated chondroitin sulfate from *H. grisea* inhibited the binding of P- and L-selectin to immobilized sialyl Lewis(x), which is a component of cell surface glycoproteins present in leukocytes and overexpressed in several tumor cells (Borsig et al. 2007). The glycan also inhibited LS180 carcinoma cell attachment to immobilized P- and L-selectins (Borsig et al. 2007). As a result of its antiselectin effect, the sea cucumber glycan attenuated lung colonization by adenocarcinoma MC-38 cells in an experimental metastasis model in mice, as well as neutrophil recruitment in thioglycollate-induced peritonitis and lipopolysaccharide-induced lung inflammation. Inhibition occurred at a dose that produces no significant change in the coagulation status of the blood. Removal of the sulfated fucose branches on the glycan abolished the inhibitory effect in vitro and in vivo (Borsig et al. 2007). These results suggest that this glycan may be a potential therapeutic drug for blocking metastasis and inflammatory reactions.

7.2.4.2 Glucosaminoglycans

In echinoderms, heparin-like glycosaminoglycans have been described only in the class Echinoidea (Table 7.2). A heparin composed by [IdoA2S–GlcNS,6S] disaccharide units, typical of mammalian heparins, with low anticoagulant activity is present exclusively in the mellitid *Mellita quinquisperforata* (sand dollar) (Demir et al. 2001; Medeiros et al. 2000).

7.2.5 *Chordata*

7.2.5.1 Galactosaminoglycans

Dermatan sulfate and chondroitin sulfate polymers with high content of disulfated disaccharide units have been described in two classes of chordate: Ascidiacea and Myxini (Table 7.1). Ascidians from the orders Phlebobranchia and Stolidobranchia contain dermatan sulfates with different disaccharide units and unique sulfation patterns. An oversulfated dermatan sulfate consisting of [IdoA2S–GalNAc6S] (80%) and [IdoA–GalNAc6S] disaccharide units has been isolated from the internal organs of the phlebobranch *Phallusia nigra* (black ascidian) (Pavao et al. 1995, 1998; Mourao et al. 1997; Vicente et al. 2001; Bao et al. 2005). Highly sulfated dermatan sulfates with the same core structure, [IdoA2–GalNAc], but sulfated at carbon 4 of the GalNAc residues, occur in the extracellular matrix of different organs (intestine, heart, mantle, and pharynx) of the stolidobranchs *Styela plicata* (sea squirt) and *Halocynthia pyriformis* (Pavao et al. 1998; Gandra et al. 2006; Hikino et al. 2003). In *S. plicata,* the 2,4-O-sulfated dermatan sulfate seems to be linked to a decorin-like core protein, which interacts with collagen (Gandra et al. 2006). No GlcAc residues were detected in the dermatan sulfates from all ascidians. Interestingly, although the dermatan sulfates from species belonging to different orders share the same sulfation pattern on the IdoA residue (2-O-sulfation), dermatans from ascidians of different orders appear to differ in the sulfation pattern of the GalNAc residue (4-O-sulfated in Stolidobranchia; 6-O-sulfation in Phlebobranchia) (Table 7.1). This fact may indicate that the expression of "specific dermatan sulfate sulfotransferases" is order specific and an evolution-related event. Although all dermatan sulfates described so far in ascidian occur in the organs of the invertebrate, recent work has detected a chondroitin sulfate in the tunic of the stolidobranchia *Styela clava* (sea squirt) (Xu et al. 2008a).

In the class Myxini, an oversulfated chondroitin sulfate/dermatan sulfate hybrid glycan, consisting of [GlcA/IdoUA–GalNAc4S,6S] disaccharide units, has been described in the notochord of the myxinid *Eptatretus burgeri* (Hagfish) (Nandini et al. 2004).

The unique oversulfated dermatan sulfates from the invertebrate chordates, found in the orders Ascidiacea and Myxini, have been extensively studied in terms of their anticoagulant/antithrombotic activity (Table 7.1) (Pavao et al. 1994, 1995, 1998; Vicente et al. 2001) and neurite outgrowth-promoting effect (Nandini et al. 2004; Hikino et al. 2003; Bao et al. 2005).

The dermatan sulfate from *S. plicata* and *H. pyriformis*, composed mainly by the disulfated disaccharide [IdoA2S–GalNAc4S], has high anticoagulant activity, which is mediated by heparin cofactor II potentiation (Pavao et al. 1998). The heparin cofactor II activities of the invertebrate glycans are about 10 times higher than that of mammalian dermatan sulfate probably due to the higher concentration of [IdoA2S–GalNAc4S]-containing sequences in the invertebrate dermatan sulfates (Pavao et al. 1998). Similar to their mammalian counterpart, the anticoagulant mechanism of the invertebrate glycans involves the induction of a covalent complex between the inhibitor and enzyme (Vicente et al. 2001). In addition, they bind to the same dermatan sulfate-binding site on the heparin cofactor II molecule as the mammalian glycan (Pavao et al. 1998). Interestingly, the dermatan sulfate obtained from the ascidian *P. nigra*, possessing the same sulfation degree but composed mainly by [IdoA2S–GalNAc6S] disaccharide units (Pavao et al. 1995), has very low anticoagulant activity and is a very weak activator of heparin cofactor II (Pavao et al. 1998), as a result of a discernible binding to the inhibitor (Vicente et al. 2001). Overall, these results indicate that binding of oversulfated dermatan sulfate polymers to heparin cofactor II requires a specific sulfation pattern on the glycans, composed by [IdoA2S–GalNAc4S]-enriched sequences.

Using these unique oversulfated dermatan sulfates from ascidians, with opposed heparin cofactor II activities, it was possible to show the relationship between heparin cofactor II activity and the antithrombotic effect. For example, the *S. plicata* dermatan sulfate with high heparin cofactor II activity prevents thrombus formation in veins after intravascular administration (Vicente et al. 2001), whereas the *P. nigra* dermatan sulfate, with a discernible heparin cofactor II activity, has no antithrombotic effect in the same venous thrombotic model, after administration of the same dose (Vicente et al. 2001). These results indicate that the antithrombotic effect of dermatan sulfate polymers involves a heparin cofactor II-mediated mechanism. This conclusion is further supported by the fact that intravascular administration of the high heparin cofactor II-active *S. plicata* dermatan sulfate to heparin cofactor II-deficient mice does not prolong the time to occlusion of the carotid artery after endothelial lesion, whereas in wild-type animals the time to occlusion was significantly prolonged (Vicente et al. 2004). However, interestingly, the relationship between heparin cofactor II activity and antithrombotic effect is not linear, since mammalian dermatan sulfate, which has an heparin cofactor II activity tenfold lower than that of the *S. plicata* glycan, has an antithrombotic effect similar to the invertebrate dermatan sulfate (Vicente et al. 2001).

The effects of the oversulfated dermatan sulfates from ascidians and Hagfish were also studied in mouse hippocampal neurons. The dermatan sulfates produced a significant neurite outgrowth-promoting activity, which resulted in distinct morphological features depending on the individual structural features of the glycans. Dermatan sulfate from *P. nigra* [IdoUA2S–GalNAc6S] induced a flattened neuronal cell soma and dendrite-like multiple neurites, whereas that from *E. burgeri* [IdoUA–GalNAc4S6S] induced a relatively small cell soma and a lower neurite sprouting with a prominent long neurite (Hikino et al. 2003). *S. plicata* dermatan sulfate [IdoUA2S–GalNAc4S], on the other hand, exhibited only a modest neurite outgrowth-promoting activity (Bao et al. 2005; Hikino et al. 2003).

7.2.5.2 Glucosaminoglycans

In marine invertebrate chordates, heparin-like glycosaminoglycans have been described only in the species *S. plicata* of stolidobranchia ascidians (Table 7.2).

Heparins composed mainly of [IdoA2S–GlcNS,6S] (major) and [HexA–GlcNS,6S] disaccharide units have been detected in intracellular granules of mast cell-like cells, present in different organs of the ascidian (intestine, heart, pharynx, mantle, and oocyte test cells) (Cavalcante et al. 2000, 2002). In the same ascidian, a heparin composed mainly by [IdoA2S–GlcNS] and [IdoA2S–GlcNS,6S] disaccharide units and lower amounts of GlcNAc3S-containing units [IdoA2S– GlcNS,3S] and [IdoA2S–GlcNS,3S,6S] has recently been detected in intracellular granules of a basophil-like cell, which circulates in the hemolymph (de Barros et al. 2007).

The effects of the *S. plicata* heparins on coagulation, thrombosis, and inflammation have been studied by experiments in vitro (Cavalcante et al. 2000; Gandra et al. 2000) and in vivo (Cardilo-Reis et al. 2006; Santos et al. 2007; Belmiro et al. 2009). The heparin isolated from oocyte test cells has an antifactor Xa activity about 50 times lower than mammalian heparin (Cardilo-Reis et al. 2006). After intravascular administration to rats (5 mg/kg), the glycan increases plasma aPTT 1.8 times and inhibits 50% of thrombosis in a venous thrombosis model without any bleeding effect. Mammalian heparin, at a dose 10 times lower, produces a fivefold increase in aPTT and inhibits 100% of thrombosis in the same model (Cardilo-Reis et al. 2006). These results are in accordance with the lower anti-Xa activity of the invertebrate heparin.

The heparin isolated from the whole body of *S. plicata* has low antithrombin activity. The invertebrate heparin has an aPTT activity of 18 IU/mg and an antithrombin-mediated antifactor IIa and antifactor Xa activities tenfold lower than that of mammalian heparin. Interestingly, in vivo experiments in rats revealed distinct effects on venous and arterial models of thrombosis. In the venous model, *S. plicata* heparin inhibits only 80% of thrombosis at a dose tenfold higher than that of the mammalian heparin that inhibits 100% of thrombosis. However, in the arterial model, both invertebrate and mammalian heparins possess equivalent antithrombotic activities (Santos et al. 2007). These results suggest that the antithrombin-mediated anticoagulant activity of heparin polymers is not directly related to antithrombotic potency in arterial thrombosis.

The effect of *S. plicata* heparin was investigated in a TNBS-induced colitis model in rats. Subcutaneous administration of the invertebrate heparin during a 7-day period drastically reduced inflammation (Belmiro et al. 2009). Heparin treatment normalized the macroscopic and histological characteristics of the colon, and, at the molecular level, a reduction of the levels of TNF-α, TGF-β, and VEGF was observed. At the cellular level, lymphocyte and macrophage recruitment and epithelial cell apoptosis were decreased. A drastic reduction in collagen-mediated fibrosis was also observed. No hemorrhagic events were observed after glycan treatment (Belmiro et al. 2009). Although the molecular mechanism is yet to be revealed, attenuation of NF-κB and MAPK kinase signaling seems to be involved in the anti-inflammatory effect of the invertebrate glycan (Belmiro et al. 2009). These results strongly indicate the potential therapeutic use of this compound for the treatment of colonic inflammation with a lower risk of hemorrhage, when compared with mammalian heparin.

7.3 Therapeutic Possibilities Involving the Marine Invertebrate Glycans

Heparin has been largely used in the clinical practice for anticoagulation and prevention/treatment of vascular thromboembolism (Roden et al. 1992; Hirsh et al. 2001). However, recent studies have shown that in addition to its well-established antithrombin-mediated thrombin and Factor-Xa inhibitory activity, heparin inhibits leukocyte recruitment (Wang et al. 2002) and tumor cell-platelet aggregation (Borsig et al. 2007) by a mechanism involving P- and L-selectin inhibition. Although heparin is currently approved only as an anticoagulant and/or antithrombotic, its potent antiselectin activity suggests that heparin could be used as an adjuvant therapeutic drug in the treatment of selectin-mediated diseases, such as inflammation and metastasis (Borsig et al. 2007; Wang et al. 2002). In fact, a recent human clinical trial has been conducted to evaluate the effects of low-molecular-weight heparin on survival and disease progression in patients with hormone-refractory prostate cancer, locally advanced pancreatic cancer, or nonsmall-cell lung carcinoma (ClinicalTrials.gov identifier: NCT00312013).

An extension of the use of heparin to other clinical conditions would increase the world's already high demand for the glycan. Heparin consumption has recently reached 100 tons per year, partially due to an increasing use of low-molecular-weight heparin (Alban 2005). It was estimated that in 2008 about 20 million people suffered from thromboembolic diseases, and 200 million pigs were needed to meet this demand (Alban 2005). Although significant progress in the chemical synthesis of complex oligosaccharides is being made (Martin et al. 2009; Seeberger and Werz 2007), the source of pharmaceutical heparins is very limited, since it is primarily obtained from porcine intestine (Liu et al. 2009). Therefore, there is an urgent necessity to find alternative sources of heparin.

In this review, several heparin analogs obtained from different marine invertebrate sources were described (Tables 7.1 and 7.2). Some of these compounds have been extensively studied in terms of structure, biological activity, and mechanism of action and evaluated in preclinical experiments in rodent animals with promising results. For example, heparin-like polymers with significant antithrombotic activity and devoid of bleeding effects occur in ascidians (Santos et al. 2007) and mollusks (Gomes et al. 2010). A unique fucosylated chondroitin sulfate possessing anticoagulant activity after oral administration (Fonseca and Mourao 2006) and high anti-metastatic (Borsig et al. 2007) and anti-inflammatory activities (Borsig et al. 2007; Belmiro et al. 2009) is present in sea cucumbers. Moreover, dermatan sulfates with significant neurite outgrowth-promoting activity occur in ascidians and sea urchins (Bao et al. 2005; Hikino et al. 2003).

The critical question related to therapeutics from natural sources is the technical and economic possibility to obtain large quantities of the compounds in a constant and ecologically correct manner. Overall, the invertebrate glycans are isolated at reasonable yields [about 0.5% of the dry weight, comparing to 0.022% from pig intestinal mucosa (Linhardt et al. 1992)] by procedures similar to those already employed in the preparation of pharmaceutical heparin. Several species of mollusks and sea cucumbers, including those containing high quantities of heparin analogs, have been successfully cultivated in different parts of the world. The cultivation employs developed aquaculture technologies capable of producing ton-quantities of starting material (Conand 2004; Helm and Bourne 2004; Lem 2005). For example, in 2001, the world's production of sea cucumber reached about 21,000 tons (Conand 2004), and that of scallops, in 1999, about 73,000 tons (Bourne 2000). Therefore, the critical conditions required to use marine invertebrates as a source of natural therapeutic compounds have already been established. What is necessary now is a cooperative effort from scientists of related areas to specifically adapt current methodologies.

Acknowledgments The work was supported by Conselho Nacional de Desenvolvimento Científico e Tecnológico (CNPq), Fundação de Amparo à Pesquisa do Estado do Rio de Janeiro (FAPERJ), Mizutani Foundation for Glycoscience (to MSGP); MSGP is a research fellow from FAPERJ and CNPq.

References

Alban S (2005) From heparins to factor xa inhibitors and beyond. Eur J Clin Invest 35(Suppl 1): 12–20

Aletras AJ, Karamanos NK, Hjerpe A (1991) Presence of the hnk-1 epitope (3-sulfoglucuronic acid) on oligosaccharides from squid skin chondroitin proteoglycans and degradation of proteoglycans by proteolytic enzymes. Biochem Int 25(2):331–338

Ariga T, Kohriyama T, Freddo L, Latov N, Saito M, Kon K, Ando S, Suzuki M, Hemling ME, Rinehart KL Jr et al (1987) Characterization of sulfated glucuronic acid containing glycolipids reacting with igm m-proteins in patients with neuropathy. J Biol Chem 262(2):848–853

Arumugam M, Balasubramanian T, Shanmugam A, Garg H (2008) A composition study of anticoagulant and antiproliferative activity of fast moving heparin from giant clam tridacna maxima (roeding, 1798) and green mussel perna viridis (Linnaeus, 1758). J Appl Biol Sci 2 (3):27–30

Bao X, Pavao MS, Dos Santos JC, Sugahara K (2005) A functional dermatan sulfate epitope containing iduronate(2-o-sulfate)alpha1-3galnac(6-o-sulfate) disaccharide in the mouse brain: demonstration using a novel monoclonal antibody raised against dermatan sulfate of ascidian *Ascidia nigra*. J Biol Chem 280(24):23184–23193

Belmiro CL, Castelo-Branco MT, Melim LM, Schanaider A, Elia C, Madi K, Pavao MS, de Souza HS (2009) Unfractionated heparin and new heparin analogues from ascidians (chordate-tunicate) ameliorate colitis in rats. J Biol Chem 284(17):11267–11278

Bergefall K, Trybala E, Johansson M, Uyama T, Naito S, Yamada S, Kitagawa H, Sugahara K, Bergstrom T (2005) Chondroitin sulfate characterized by the e-disaccharide unit is a potent inhibitor of herpes simplex virus infectivity and provides the virus binding sites on gro2c cells. J Biol Chem 280(37):32193–32199

Borsig L, Wang L, Cavalcante MC, Cardilo-Reis L, Ferreira PL, Mourao PA, Esko JD, Pavao MS (2007) Selectin blocking activity of a fucosylated chondroitin sulfate glycosaminoglycan from sea cucumber. Effect on tumor metastasis and neutrophil recruitment. J Biol Chem 282 (20):14984–14991

Bourne NF (2000) The potential for scallop culture – the next millenium. Aquac Int 8(2–3): 113–122

Brito AS, Arimateia DS, Souza LR, Lima MA, Santos VO, Medeiros VP, Ferreira PA, Silva RA, Ferreira CV, Justo GZ, Leite EL, Andrade GP, Oliveira FW, Nader HB, Chavante SF (2008) Anti-inflammatory properties of a heparin-like glycosaminoglycan with reduced anti-coagulant activity isolated from a marine shrimp. Bioorg Med Chem 16(21):9588–9595

Burson SL, Fahrenbach MJ, Frommhagen LH, Riccardi BA, Brown RA, Brockman JA, Lewry HV, Stokstad ELR (1956) Isolation and purification of mactins, heparin-like anticoagulants from mollusca. J Am Chem Soc 78(22):5874–5878

Buyue Y, Sheehan JP (2009) Fucosylated chondroitin sulfate inhibits plasma thrombin generation via targeting of the factor ixa heparin-binding exosite. Blood 114(14):3092–3100

Cardilo-Reis L, Cavalcante MC, Silveira CB, Pavao MS (2006) In vivo antithrombotic properties of a heparin from the oocyte test cells of the sea squirt styela plicata(chordata-tunicata). Braz J Med Biol Res 39(11):1409–1415

Cassaro CM, Dietrich CP (1977) Distribution of sulfated mucopolysaccharides in invertebrates. J Biol Chem 252(7):2254–2261

Cavalcante MC, Allodi S, Valente AP, Straus AH, Takahashi HK, Mourao PA, Pavao MS (2000) Occurrence of heparin in the invertebrate styela plicata (tunicata) is restricted to cell layers facing the outside environment. An ancient role in defense? J Biol Chem 275(46): 36186–36189

Cavalcante MC, de Andrade LR, Du Bocage Santos-Pinto C, Straus AH, Takahashi HK, Allodi S, Pavao MS (2002) Colocalization of heparin and histamine in the intracellular granules of test cells from the invertebrate styela plicata (chordata-tunicata). J Struct Biol 137(3):313–321

Cesaretti M, Luppi E, Maccari F, Volpi N (2004) Isolation and characterization of a heparin with high anticoagulant activity from the clam tapes phylippinarum: evidence for the presence of a high content of antithrombin iii binding site. Glycobiology 14(12):1275–1284

Chavante SF, Santos EA, Oliveira FW, Guerrini M, Torri G, Casu B, Dietrich CP, Nader HB (2000) A novel heparan sulphate with high degree of n-sulphation and high heparin cofactor-II activity from the brine shrimp artemia franciscana. Int J Biol Macromol 27(1):49–57

Chou DK, Ilyas AA, Evans JE, Costello C, Quarles RH, Jungalwala FB (1986) Structure of sulfated glucuronyl glycolipids in the nervous system reacting with hnk-1 antibody and some igm paraproteins in neuropathy. J Biol Chem 261(25):11717–11725

Clement AM, Sugahara K, Faissner A (1999) Chondroitin sulfate e promotes neurite outgrowth of rat embryonic day 18 hippocampal neurons. Neurosci Lett 269(3):125–128

Conand C (2004) Present status of world sea cucumber resoruces and utilisation: an international overview. Advances in sea cucumber aquaculture and management. FAO, Rome

de Barros CM, Andrade LR, Allodi S, Viskov C, Mourier PA, Cavalcante MC, Straus AH, Takahashi HK, Pomin VH, Carvalho VF, Martins MA, Pavao MS (2007) The hemolymph of the ascidian styela plicata (chordata-tunicata) contains heparin inside basophil-like cells and a unique sulfated galactoglucan in the plasma. J Biol Chem 282(3):1615–1626

Deepa SS, Umehara Y, Higashiyama S, Itoh N, Sugahara K (2002) Specific molecular interactions of oversulfated chondroitin sulfate e with various heparin-binding growth factors. Implications as a physiological binding partner in the brain and other tissues. J Biol Chem 277(46): 43707–43716

Demir M, Iqbal O, Dietrich CP, Hoppensteadt DA, Ahmad S, Daud AN, Fareed J (2001) Anticoagulant and antiprotease effects of a novel heparinlike compound from shrimp (penaeus brasiliensis) and its neutralization by heparinase i. Clin Appl Thromb Hemost 7(1):44–52

Dietrich CP, de Paiva JF, Moraes CT, Takahashi HK, Porcionatto MA, Nader HB (1985) Isolation and characterization of a heparin with high anticoagulant activity from Anomalocardia brasiliana. Biochim Biophys Acta 843(1–2):1–7

Dietrich CP, Nader HB, de Paiva JF, Santos EA, Holme KR, Perlin AS (1989) Heparin in molluscs: chemical, enzymatic degradation and 13C and 1H N.M.R. Spectroscopical evidence for the maintenance of the structure through evolution. Int J Biol Macromol 11(6):361–366

Dietrich CP, Paiva JF, Castro RA, Chavante SF, Jeske W, Fareed J, Gorin PA, Mendes A, Nader HB (1999) Structural features and anticoagulant activities of a novel natural low molecular weight heparin from the shrimp *Penaeus brasiliensis*. Biochim Biophys Acta 1428(2–3): 273–283

Dreyfuss JL, Regatieri CV, Lima MA, Paredes-Gamero EJ, Brito AS, Chavante SF, Belfort R, Farah ME, Nader HB (2010) A heparin mimetic isolated from a marine shrimp suppresses neovascularization. J Thromb Haemost 8:1828–1837

Feta A, Do AT, Rentzsch F, Technau U, Kusche-Gullberg M (2009) Molecular analysis of heparan sulfate biosynthetic enzyme machinery and characterization of heparan sulfate structure in *Nematostella vectensis*. Biochem J 419(3):585–593. doi:10.1042/BJ20082081

Fongmoon D, Shetty AK, Basappa YS, Sugiura M, Kongtawelert P, Sugahara K (2007) Chondroitinase-mediated degradation of rare 3-o-sulfated glucuronic acid in functional oversulfated chondroitin sulfate k and e. J Biol Chem 282(51):36895–36904

Fonseca RJ, Mourao PA (2006) Fucosylated chondroitin sulfate as a new oral antithrombotic agent. Thromb Haemost 96(6):822–829. doi:06120822 [pii]

Gandra M, Cavalcante M, Pavao M (2000) Anticoagulant sulfated glycosaminoglycans in the tissues of the primitive chordate styela plicata (tunicata). Glycobiology 10(12):1333–1340

Gandra M, Kozlowski EO, Andrade LR, de Barros CM, Pascarelli BM, Takiya CM, Pavao MS (2006) Collagen colocalizes with a protein containing a decorin-specific peptide in the tissues of the ascidian styela plicata. Comp Biochem Physiol B Biochem Mol Biol 144(2):215–222

Glauser BF, Pereira MS, Monteiro RQ, Mourao PA (2008) Serpin-independent anticoagulant activity of a fucosylated chondroitin sulfate. Thromb Haemost 100(3):420–428

Gomes AM, Kozlowski EO, Pomin VH, de Barros CM, Zaganeli JL, Pavao MS (2010) Unique extracellular matrix heparan sulfate from the bivalve nodipecten nodosus (Linnaeus, 1758) safely inhibits arterial thrombosis after photochemically induced endothelial lesion. J Biol Chem 285(10):7312–7323

Habuchi O, Sugiura K, Kawai N (1977) Glucose branches in chondroitin sulfates from squid cartilage. J Biol Chem 252(13):4570–4576

Habuchi O, Yamagata T, Suzuki S (1971) Biosynthesis of the acetylgalactosamine 4, 6-disulfate unit of squid chondroitin sulfate by transsulfation from 3′-phosphoadenosine 5′-phosphosulfate. J Biol Chem 246(23):7357–7365

Haefner B (2003) Drugs from the deep: marine natural products as drug candidates. Drug Discov Today 8(12):536–544

Helm MM, Bourne N (2004) The hatchery culture of bivalves: a practical manual Hatchery culture of bivalves. FAO, Rome

Hikino M, Mikami T, Faissner A, Vilela-Silva AC, Pavao MS, Sugahara K (2003) Oversulfated dermatan sulfate exhibits neurite outgrowth-promoting activity toward embryonic mouse hippocampal neurons: implications of dermatan sulfate in neuritogenesis in the brain. J Biol Chem 278(44):43744–43754

Hirsh J, Warkentin TE, Shaughnessy SG, Anand SS, Halperin JL, Raschke R, Granger C, Ohman EM, Dalen JE (2001) Heparin and low-molecular-weight heparin: mechanisms of action, pharmacokinetics, dosing, monitoring, efficacy, and safety. Chest 119(1 Suppl):64S–94S

Hovingh P, Linker A (1982) An unusual heparan sulfate isolated from lobsters (homarus americanus). J Biol Chem 257(16):9840–9844

Jordan RE, Marcum JA (1986) Anticoagulantly active heparin from clam (mercenaria mercenaria). Arch Biochem Biophys 248(2):690–695

Karamanos NK (1992) Two squid skin proteoglycans each containing chondroitin sulfates with different sulfation patterns. Biochem Cell Biol 70(8):629–635

Karamanos NK, Aletras AJ, Antonopoulos CA, Hjerpe A (1994) Determination of the hnk-1 epitope (3-sulphated glucuronic acid) in intact chondroitin sulphates by ELISA. Application to squid skin proteoglycans and their oversulphated carbohydrate structures. Biochimie 76(1): 79–82

Karamanos NK, Aletras AJ, Antonopoulos CA, Tsegenidis T, Tsiganos CP, Vynios DH (1988) Extraction and fractionation of proteoglycans from squid skin. Biochim Biophys Acta 966 (1):36–43

Karamanos NK, Manouras A, Tsegenidis T, Antonopoulos CA (1991) Isolation and chemical study of the glycosaminoglycans from squid cornea. Int J Biochem 23(1):67–72

Kariya Y, Sakai T, Kaneko T, Suzuki K, Kyogashima M (2002) Enhancement of t-pa-mediated plasminogen activation by partially defucosylated glycosaminoglycans from the sea cucumber *Stichopus japonicus*. J Biochem 132(2):335–343

Kariya Y, Watabe S, Hashimoto K, Yoshida K (1990) Occurrence of chondroitin sulfate e in glycosaminoglycan isolated from the body wall of sea cucumber *Stichopus japonicus*. J Biol Chem 265(9):5081–5085

Kariya Y, Watabe S, Kyogashima M, Ishihara M, Ishii T (1997) Structure of fucose branches in the glycosaminoglycan from the body wall of the sea cucumber *Stichopus japonicus*. Carbohydr Res 297(3):273–279

Kawai Y, Seno N, Anno K (1966) Chondroitin polysulfate of squid cartilage. J Biochem 60 (3):317–321

Kawashima H, Atarashi K, Hirose M, Hirose J, Yamada S, Sugahara K, Miyasaka M (2002) Oversulfated chondroitin/dermatan sulfates containing glcabeta1/idoaalpha1–3galnac(4, 6-o-disulfate) interact with l- and p-selectin and chemokines. J Biol Chem 277(15):12921–12930

Kinoshita A, Yamada S, Haslam SM, Morris HR, Dell A, Sugahara K (1997) Novel tetrasaccharides isolated from squid cartilage chondroitin sulfate e contain unusual sulfated disaccharide units glca(3-o-sulfate)beta1-3galnac(6-o-sulfate) or glca(3-o-sulfate)beta1-3galnac. J Biol Chem 272(32):19656–19665

Kinoshita-Toyoda A, Yamada S, Haslam SM, Khoo KH, Sugiura M, Morris HR, Dell A, Sugahara K (2004) Structural determination of five novel tetrasaccharides containing 3-o-sulfated d-glucuronic acid and two rare oligosaccharides containing a beta-d-glucose branch isolated from squid cartilage chondroitin sulfate e. Biochemistry 43(34):11063–11074

Lem A (2005) International trade in aquaculture products. Topics fact sheets. Department FFaA, Rome

Lindahl U, Backstrom G, Thunberg L, Leder IG (1980) Evidence for a 3-o-sulfated d-glucosamine residue in the antithrombin-binding sequence of heparin. Proc Natl Acad Sci USA 77(11): 6551–6555

Linhardt RJ, Ampofo SA, Fareed J, Hoppensteadt D, Mulliken JB, Folkman J (1992) Isolation and characterization of human heparin. Biochemistry 31(49):12441–12445

Liu H, Zhang Z, Linhardt RJ (2009) Lessons learned from the contamination of heparin. Nat Prod Rep 26(3):313–321

Luppi E, Cesaretti M, Volpi N (2005) Purification and characterization of heparin from the Italian clam callista chione. Biomacromolecules 6(3):1672–1678

Martin JG, Gupta M, Xu Y, Akella S, Liu J, Dordick JS, Linhardt RJ (2009) Toward an artificial golgi: redesigning the biological activities of heparan sulfate on a digital microfluidic chip. J Am Chem Soc 131(31):11041–11048

Mathews MB, Duh J, Person P (1962) Acid mucopolysaccharides of invertebrate cartilage. Nature 193:378–379

Medeiros GF, Mendes A, Castro RA, Bau EC, Nader HB, Dietrich CP (2000) Distribution of sulfated glycosaminoglycans in the animal kingdom: widespread occurrence of heparin-like compounds in invertebrates. Biochim Biophys Acta 1475(3):287–294

Molinski T (2009) Marine natural products. Clin Adv Hematol Oncol 7(6):383–385

Mourao PA, Boisson-Vidal C, Tapon-Bretaudiere J, Drouet B, Bros A, Fischer A (2001) Inactivation of thrombin by a fucosylated chondroitin sulfate from echinoderm. Thromb Res 102(2): 167–176

Mourao PA, Pavao MS, Mulloy B, Wait R (1997) Chondroitin abc lyase digestion of an ascidian dermatan sulfate. Occurrence of unusual 6-o-sulfo-2-acetamido-2-deoxy-3-o-(2-o-sulfo-alpha-l-idopyranosyluronic acid)-beta-d-galactose units. Carbohydr Res 300(4):315–321

Mourao PA, Pereira MS, Pavao MS, Mulloy B, Tollefsen DM, Mowinckel MC, Abildgaard U (1996) Structure and anticoagulant activity of a fucosylated chondroitin sulfate from echinoderm. Sulfated fucose branches on the polysaccharide account for its high anticoagulant action. J Biol Chem 271(39):23973–23984

Nader HB, Ferreira TM, Paiva JF, Medeiros MG, Jeronimo SM, Paiva VM, Dietrich CP (1984) Isolation and structural studies of heparan sulfates and chondroitin sulfates from three species of molluscs. J Biol Chem 259(3):1431–1435

Nader HB, Pinhal MA, Bau EC, Castro RA, Medeiros GF, Chavante SF, Leite EL, Trindade ES, Shinjo SK, Rocha HA, Tersariol IL, Mendes A, Dietrich CP (2001) Development of new heparin-like compounds and other antithrombotic drugs and their interaction with vascular endothelial cells. Braz J Med Biol Res 34(6):699–709

Nandini CD, Mikami T, Ohta M, Itoh N, Akiyama-Nambu F, Sugahara K (2004) Structural and functional characterization of oversulfated chondroitin sulfate/dermatan sulfate hybrid chains from the notochord of hagfish. Neuritogenic and binding activities for growth factors and neurotrophic factors. J Biol Chem 279(49):50799–50809

Nobuko S, Murakami K (1982) Structure of disulfated disaccharides from chondroitin polysulfates, chondroitin sulfate d and k. Carbohydr Res 103(1):190–194

Oguri K, Yamagata T (1978) Appearance of a proteoglycan in developing sea urchin embryos. Biochim Biophys Acta 541(3):385–393. doi:0304-4165(78)90197-6 [pii]

Pavao MS, Aiello KR, Werneck CC, Silva LC, Valente AP, Mulloy B, Colwell NS, Tollefsen DM, Mourao PA (1998) Highly sulfated dermatan sulfates from ascidians. Structure versus anticoagulant activity of these glycosaminoglycans. J Biol Chem 273(43):27848–27857

Pavao MS, Mourao PA, Mulloy B, Tollefsen DM (1995) A unique dermatan sulfate-like glycosaminoglycan from ascidian. Its structure and the effect of its unusual sulfation pattern on anticoagulant activity. J Biol Chem 270(52):31027–31036

Pavao MS, Rodrigues MA, Mourao PA (1994) Acidic polysaccharides of the ascidian styela plicata. Biosynthetic studies on the sulfated l-galactans of the tunic, and preliminary characterization of a dermatan sulfate-like polymer in body tissues. Biochim Biophys Acta 1199(3):229–237

Pejler G, Danielsson A, Bjork I, Lindahl U, Nader HB, Dietrich CP (1987) Structure and antithrombin-binding properties of heparin isolated from the clams anomalocardia brasiliana and tivela mactroides. J Biol Chem 262(24):11413–11421

Pomin VH (2009) Review: an overview about the structure-function relationship of marine sulfated homopolysaccharides with regular chemical structures. Biopolymers 91(8):601–609

Pomin VH, Mourao PA (2008) Structure, biology, evolution, and medical importance of sulfated fucans and galactans. Glycobiology 18(12):1016–1027

Roden L, Ananth S, Campbell P, Curenton T, Ekborg G, Manzella S, Pillion D, Meezan E (1992) Heparin – an introduction. Adv Exp Med Biol 313:1–20

Santos EA, Rocha LR, Pereira NM, Andrade GP, Nader HB, Dietrich CP (2002) Mast cells are present in epithelial layers of different tissues of the mollusc anomalocardia brasiliana. In situ characterization of heparin and a correlation of heparin and histamine concentration. Histochem J 34(11–12):553–558

Santos JC, Mesquita JM, Belmiro CL, da Silveira CB, Viskov C, Mourier PA, Pavao MS (2007) Isolation and characterization of a heparin with low antithrombin activity from the body of styela plicata (chordata-tunicata). Distinct effects on venous and arterial models of thrombosis. Thromb Res 121(2):213–223

Saravanan R, Shanmugam A (2010a) Isolation and characterization of low molecular weight glycosaminoglycans from marine mollusc amussium pleuronectus (linne) using chromatography. Appl Biochem Biotechnol 160(3):791–799

Saravanan R, Shanmugam A (2010b) Preventive effect of low molecular weight glycosaminoglycan from amussium pleuronectus (linne) on oxidative injury and cellular abnormalities in isoproterenol-induced cardiotoxicity in wistar rats. Appl Biochem Biotechnol 162(1):43–51

Seeberger PH, Werz DB (2007) Synthesis and medical applications of oligosaccharides. Nature 446(7139):1046–1051

Shetty AK, Kobayashi T, Mizumoto S, Narumi M, Kudo Y, Yamada S, Sugahara K (2009) Isolation and characterization of a novel chondroitin sulfate from squid liver integument rich in n-acetylgalactosamine(4, 6-disulfate) and glucuronate(3-sulfate) residues. Carbohydr Res 344(12):1526–1532

Sugahara K, Tanaka Y, Yamada S, Seno N, Kitagawa H, Haslam SM, Morris HR, Dell A (1996) Novel sulfated oligosaccharides containing 3-o-sulfated glucuronic acid from king crab cartilage chondroitin sulfate k. Unexpected degradation by chondroitinase abc. J Biol Chem 271 (43):26745–26754

Tamura J, Arima K, Imazu A, Tsutsumishita N, Fujita H, Yamane M, Matsumi Y (2009) Sulfation patterns and the amounts of chondroitin sulfate in the diamond squid, *Thysanoteuthis rhombus*. Biosci Biotechnol Biochem 73(6):1387–1391

Tapon-Bretaudiere J, Drouet B, Matou S, Mourao PA, Bros A, Letourneur D, Fischer AM (2000) Modulation of vascular human endothelial and rat smooth muscle cell growth by a fucosylated chondroitin sulfate from echinoderm. Thromb Haemost 84(2):332–337

Ueoka C, Kaneda N, Okazaki I, Nadanaka S, Muramatsu T, Sugahara K (2000) Neuronal cell adhesion, mediated by the heparin-binding neuroregulatory factor midkine, is specifically inhibited by chondroitin sulfate e. Structural and functional implications of the over-sulfated chondroitin sulfate. J Biol Chem 275(48):37407–37413

Ulrich PN, Boon JK (2001) The histological localization of heparin in the northern quahog clam, mercenaria mercenaria. J Invertebr Pathol 78(3):155–159

Vicente CP, He L, Pavao MS, Tollefsen DM (2004) Antithrombotic activity of dermatan sulfate in heparin cofactor II-deficient mice. Blood 104(13):3965–3970

Vicente CP, Zancan P, Peixoto LL, Alves-Sa R, Araujo FS, Mourao PA, Pavao MS (2001) Unbalanced effects of dermatan sulfates with different sulfation patterns on coagulation, thrombosis and bleeding. Thromb Haemost 86(5):1215–1220

Vieira RP, Mourao PA (1988) Occurrence of a unique fucose-branched chondroitin sulfate in the body wall of a sea cucumber. J Biol Chem 263(34):18176–18183

Vieira RP, Mulloy B, Mourao PA (1991) Structure of a fucose-branched chondroitin sulfate from sea cucumber. Evidence for the presence of 3-o-sulfo-beta-d-glucuronosyl residues. J Biol Chem 266(21):13530–13536

Vijayabaskar P, Balasubramanian T, Somasundaram ST (2008) Low-molecular weight molluscan glycosaminoglycan from bivalve katelysia opima (gmelin). Methods Find Exp Clin Pharmacol 30(3):175–180

Vijayabaskar P, Somasundaram ST (2009) Histological localization and analytical evaluation of heparin in the wedge clam donax cuneatus (linnaeus, 1758). Method Find Exp Clin 31(5): 311–316

Vilela-Silva AC, Werneck CC, Valente AP, Vacquier VD, Mourao PA (2001) Embryos of the sea urchin strongylocentrotus purpuratus synthesize a dermatan sulfate enriched in 4-o- and 6-o-disulfated galactosamine units. Glycobiology 11(6):433–440

Volpi N, Maccari F (2009) Structural characterization and antithrombin activity of dermatan sulfate purified from marine clam scapharca inaequivalvis. Glycobiology 19(4):356–367

Wang L, Brown JR, Varki A, Esko JD (2002) Heparin's anti-inflammatory effects require glucosamine 6-o-sulfation and are mediated by blockade of l- and p-selectins. J Clin Invest 110(1):127–136

Xu CX, Jin H, Chung YS, Shin JY, Lee KH, Beck GR Jr, Palmos GN, Choi BD, Cho MH (2008a) Chondroitin sulfate extracted from ascidian tunic inhibits phorbol ester-induced expression of inflammatory factors vcam-1 and cox-2 by blocking nf-kappab activation in mouse skin. J Agric Food Chem 56(20):9667–9675

Xu CX, Jin H, Chung YS, Shin JY, Woo MA, Lee KH, Palmos GN, Choi BD, Cho MH (2008b) Chondroitin sulfate extracted from the styela clava tunic suppresses tnf-alpha-induced expression of inflammatory factors, vcam-1 and inos by blocking akt/nf-kappab signal in jb6 cells. Cancer Lett 264(1):93–100

Yamada S, Morimoto H, Fujisawa T, Sugahara K (2007) Glycosaminoglycans in hydra magnipapillata (hydrozoa, cnidaria): demonstration of chondroitin in the developing nematocyst, the sting organelle, and structural characterization of glycosaminoglycans. Glycobiology 17 (8):886–894

Yamaguchi M, Kinoshita S, Suzuki N (1989) Dermatan sulfate formation in gastrulae of the sea urchin *Clypeaster japonicus*. J Biochem 106(1):158–162

Zancan P, Mourao PA (2004) Venous and arterial thrombosis in rat models: dissociation of the antithrombotic effects of glycosaminoglycans. Blood Coagul Fibrinolysis 15(1):45–54

Index

A

N-Acetylgalactosamine, 88
N-Acetylglucosamine, 5, 88
N-Acetylglucosamine
  N-deacetylase–N-sulfotransferases
  (Ndst), 8
  inactivation in mice, 8–18, 117–121
  in endothelial cells, 8
  role in heparan sulfate modification, 8
ADAM17, 33
Adiponectin, 94
Agrin, 32
Angiogenesis, 18–24
  defined, 18
  inhibitors of, 151
  role of heparan sulfate in, 18–24
  tumor, 19, 147–148
Anticoagulant activity
  dependent on sulfation, 6–7
  of heparin, 5–8
Apoptosis, 143–144
Arthopoda, GAGs in, 170–171
Atherosclerosis, 83–104
  pathogenesis, 85–87
  scheme of early atherosclerosis, 86
  proteoglycans in, 87–95

B

Basement membrane, 32–24, 83
Biglycan, 86, 93–94
  knockout mice, 94

C

Cancer
  alteration in glycosylation, 65 (table), 64–76
  glycans in cancer, 63–76, 135–152
  glycosylation and metastasis, 74
  therapy, 148–151
  tumor angiogenesis, 19–20
CD44, 95
Chemokines, 14
  binding to heparan sulfate, 14–15
  gradients, 17–18
  transcytosis, 15
Chondroitin sulfate proteoglycan, 86
  in atherosclerosis, 92–94
  sugar composition, 88
Chordata, GAGs in, 175–177
Cnidaria, GAGs in, 169–170

D

Decorin, 93–94
  in thrombosis, 102
Dengue virus, 41
Dermatan sulfate proteoglycan, 86
  in atherosclerosis, 92–94
  sugar composition, 88
  thrombosis, 99–101
Diabetes, 87

E

Ecinodermata, GAGs in, 172–174
Elastic fibers, 84
Elastic lamina, 84
Endocytosis
  mediated by HSPGs, 34
Endothelial cell
  heparan sulfate on, 8–24
  in blood vessels, 83–85
Epimerisation, 118
ES cells, *See* stem cells
Ext HS co-polymerase genes, 117–121

M.S.G. Pavão (ed.), *Glycans in Diseases and Therapeutics*, Biology of Extracellular Matrix,
DOI 10.1007/978-3-642-16833-8,

F
Fibulin, 93

G
GAG, *See* glycosaminoglycan
GalNac, *See* N-acetylgalactosamine
GlcA, *See* D-glucuronic acid
GlcNAc, *See* N-acetylglucosamine
GlcNH2, *See* glucosamine
GlcNS, *See* N-sulfoglucosamine
Glycosaminoglycan, 32
Glucosamine, 5, 88
D-Glucuronic acid, 1, 88
Glycan synthesis, 64
Glycosaminoglycan, 113–114
  in atherosclerosis, 83–104
  in marine invertebrates, 159–179
  in thrombosis, 95–103
Glycosylation
  altered in cancer, 64–76
  metastasis, 74
  N-linked, 66–69
  O-linked, 69–73
Glypican, 32–34

H
HAS enzymes, 95
Heparan sulfate, 1–24,32–52
  angiogenesis, 18–24, 147–148, 151
  atherosclerosis, 90–92
  chemokine binding and presentation, 14–15
  endothelial cells, 8–24
  knockout mouse, 117–121
  growth factor binding, 116
  in cancer therapy, 148–151
  infection, 31–52
  inflammation, 1–3, 7–24
  inhibitors of biosynthesis, 149–151
  interactions with PECAM-1, 18–24
  interactions with VEGF, 18–24
  knockout mice, 8–18
  leukocyte interactions, 11–13
  primer on biology, 32–34
  thrombosis, 98–99
  transcytosis of chemokines, 15–17
Heparan sulfate proteoglycan, 32–54
  angiogenesis, 18–24, 147–148, 151
  cancer, 135–152
  cell motility, 144–145
  function, 114
  infection, 32–54
  inflammation, 1–3, 7–24
  mesodermal differentiation of ES cells, 128–130
  microbial attachment and internalization, 35–42
  neuroectodermal differentiation of stem cells, 126–128
  receptor for virulence factors, 42–44
  pathogen interactions, 36–37 (Table), 35–42
  peptide-based inhibition of HSPG function, 150
  primer on biology, 32–34
  stem cell biology, 113–131
  structure and biosynthesis, 114–116
  sugar composition, 88
  tumor progression, 141–148
Heparanase, 143–147
Heparin
  affect on inflammation, 5–7
  anticoagulant activity, 5–7
  biological function, 1–2
  inhibition of selectin function, 3–7
  sugar composition, 5–6, 88
  thrombosis, 97–98
HIV, 42–44
HS, *See* heparan sulfate
HS endosulfatase (HSulf), 143
HSepi gene knockout, 118
HSPG, *See* heparan sulfate proteoglycan
Hs2st gene knockout, 119
Hs6st gene knockout, 119
Hyaluronic acid, 95
Hyperglycemia, 87

I
ICAM, 14
IdoA, *See* L-iduronic acid
L-Iduronic acid, 1, 88
IL8, 15–17
Infection
  role of heparan sulfate proteoglycans, 31–52
Inflammation, 2–18
  function of heparin, 2–18
  leukocyte trafficking, 2–3, 17–18
  role of heparan sulfate, 2–18
Integrin, 143–147
Intimatan, 102
Invertebrate glycans, *See* marine invertebrates

K
KC, 14–15

L
Leukocyte
  heparan sulfate on, 8–10
  neutrophil crawling, 17–18

role of selectins in recruitment, 3–5, 11–13
trafficking in inflammation, 2–3
Listeria monocytogenes, 42

M
Malaria, 44–45
Marine invertebrates, 159–179
Arthopoda, GAGs in, 170–171
Chordata, GAGs in, 175–177
Cnidaria, GAGs in, 169–170
Ecinodermata, GAGs in, 172–174
GAGs and biological properties, 161–169
Mollusca, GAGs in, 171–172
therapeutic possibilities Mast cells, 2
Matrilysin, *See* MMP7
Matrix metalloproteinases
in cancer, 146–147
syndecan shedding, 33–34
Metastasis, 74
Microbe attachment and internalization to HSPG, 35–42
dengue virus, 41
Listeria monocytogenes, , 42
papillomavirus, 38–41
subversion of HSPG shedding, 50
MIP2, 14
MMP, *See* matrix metalloproteinase
MMP2
in metastasis, 146
MMP7
role in syndecan shedding, 33
MMP9
in cancer, 146–147
role in syndecan shedding, 33
MMP14
role in syndecan shedding, 33
Mollusca, GAGs in, 171–172
mSulf gene knockouts, 119
MUC1–4, 72–73

N
Ndst, *See* N-acetylglucosamine N-deacetylase–N-sulfotransferases
Neutrophil, *See* leukocyte

O
ost gene knockouts, 118–121

P
Papillomavirus, 38–41
PECAM-1, 18–24
Perlecan, 32–34
Plasmodium falciparum, 44–45
Pseudomonas aeruginosa, 47–48

S
Selectins, 3, 4–24, 75–76
in tumor metastasis, 74–76
inhibition by heparin, 3–5
knockout mice, 4
recognition of sialyl Lewis^X, 3
role in leukocyte recruitment, 3–5
Shedding
in bacterial pathogenesis, 47–51
of Glypican, 34
of syndecan ectodomains, 33–34
Sialyl Lewis^X (SLe^X), 3
Stem Cells, 113–131
culturing mouse and human stem cells, 123–125
deriving and defining embryonic stem cells, 121–122
mesodermal differentiation, 128–131
neuroectodermal differentiation, 126–128
pluripotency and self renewal, 122–123
Staphylococcus aureus, 48–50
Streptococcus pneumonia, 50–51
Sulfated tyrosine, 3
Sulfation of carbohydrate
N-sulfation, 117–118
6-O-desulfation, 119
2-O-sulfation, 118
3-O-sulfation, 119–120
6-O-sulfation, 118–119
N-Sulfoglucosamine, 5
Syndecan, 32–34
apoptosis, 143
cytoplasmic domain interactions, 32–34
gene knockdown, 143
shedding, 33–34, 47–51
Synthesis of glycans, 64

T
T. gondii, 45–47
TACE, *See* ADAM17
Thioglycollate-induced peritonitis, 5
Thrombosis, 95–104
TNFα, 14
Transcytosis of chemokines, 15–17
Tumor
angiogenesis, 19–20, 147–148
apoptosis, 143–144
heparan sulfate in cancer therapy, 148–151
progression, 141–148

V
Vascular system
  smooth muscle cells, 83–85
  structure, 83–85
VCAM, 14
VEGF, 18–24
Versican, 93–94
Virulence factors, 42–44